Praise for Designing Bots

"As the world moves to conversational interfaces, everyone working in consumer or enterprise software will benefit from reading this book. Amir Shevat has written a must-have addition to every designer's bookshelf."

NIR EYAL, BESTSELLING AUTHOR OF *HOOKED: HOW TO BUILD HABIT-FORMING PRODUCTS*

"Amir is the original "Botstar," the first public persona to emerge in this new era of messaging and conversational interfaces. No one has done more than him to champion and educate the world about this nascent field and with the publication of Designing Bots he's now literally written the book on it. At Automat we've been working on conversational software for more than 15 years, even inventing the sub-category of Conversational Marketing and we still learned a ton from this book—it's now required reading for anyone who joins our company!"

ANDY MAURO, CEO OF AUTOMAT

"With his magnetic talent for bringing the bot developer community together, in person and through this wonderful book–Amir shares his experience via countless conversations, experiments and lessons learned building bots that work (or don't work) in real teams. Thank you Amir, for the brilliant insights and your commitment to making software more personal, human and adapt to the way people work and share."

LILI CHENG, AI & RESEARCH MICROSOFT

"Amir is indisputably one of the most important voices in the world of bots, informing best practices for hundreds of companies. From ethics to design, his knowledge and expertise helps to drive this new industry forward. This is a must-read for anyone who wants to build a business around bots."

VERONICA BELMONT, PRODUCT MANAGER AT GROWBOT.IO, ADMIN AT BOTWIKI.ORG

Designing Bots

Creating Conversational Experiences

Amir Shevat

Beijing · Boston · Farnham · Sebastopol · Tokyo

Designing Bots
by Amir Shevat

Printed in the United States of America.

Published by O'Reilly Media, Inc., 1005 Gravenstein Highway North, Sebastopol, CA 95472.

O'Reilly books may be purchased for educational, business, or sales promotional use. Online editions are also available for most titles (*http://oreilly.com/safari*). For more information, contact our corporate/institutional sales department: (800) 998-9938 or *corporate@oreilly.com*.

Acquisitions Editor: Mary Treseler
Editor: Angela Rufino
Production Editor: Nicholas Adams
Copyeditor: Rachel Head
Proofreader: Molly Ives Brower
Indexer: Lucie Haskins
Cover Designer: Randy Comer
Interior Designers: Ron Bilodeau and Monica Kamsvaag
Illustrator: Rebecca Demarest
Compositor: Nicholas Adams

May 2017: First Edition.

Revision History for the Second Edition:

2015-05-05 First release

See *http://oreilly.com/catalog/errata.csp?isbn= 0636920057741* for release details.

978-1-491-97482-7

[LSI]

[*contents*]

Preface

As Slack's head of developer relations, I spend my days helping people build bots: from big partners like SAP and IBM, all the way to the lone developer anywhere in the world. In this book, I hope to share with you some of the lessons I have learned.

How are bots changing the world? This happens to me every week:

> *My meeting counterpart*: Your personal assistant, Amy, is so amazing! How did you find someone so responsive?
>
> *Me*: Amy is not a human—it is software.
>
> *My meeting counterpart*: 😳

This new user interface is revolutionizing the way we interact with software. Finally, software is able to meet humans on our playing field and interact with us in an intuitive way—this blows people's minds in a way that reminds me of the first time I saw an iPhone touchscreen.

Whether you are designing a new consumer service, an enterprise product, or any other software, you should think about your conversational interface. In the future, we will build for the web, we will build for mobile apps, and we will build for conversational apps.

Designing bots is a new design proficiency, and is not a trivial matter. While bots are a great new user interface, they are not suited for every use case, and you will need to learn how to use bots effectively. There are also many considerations that need to be taken into account, from defining a core purpose to working out an effective onboarding process, outlining different flows, defining a personality, and choosing the right balance of rich control and text.

This book will take you on a journey of understanding bots and learning how to design bots. You will also read about a lot of bot designers and developers who have shared their successes, failures, and best practices. This book is practical as well as theoretical, and we will actually go through all the steps of designing two bots: a consumer bot on Facebook Messenger and a business bot on Slack.

Who Should Read This Book?

This book focuses on the design of a conversational user interface, but also covers considerations such as distribution, marketing, architecture, and monetization. If you are considering building a bot or learning how to design a conversational interface, this book is for you.

DESIGNERS

This book will be your toolkit for designing bots, a book you can go back to when tackling your own bot design. We will go into the details of the design process, and give concrete examples of designing bots for B2B and B2C use cases. We will go over everything from use case specification to actual designing of a bot, all the way to validating our design with users.

PRODUCT MANAGERS

This book will arm you with a good understanding of which use cases are better suited for bots and how you, as a product manager, can use bots to expose and extend your products. We will provide examples of how to write the specification for a good conversational interface, and share best practices on product decisions that have led to user delight.

ENTREPRENEURS

This book will give you an overview of the bot ecosystem, the opportunities that it offers to you, the engagement and monetization models this interface facilitates, and the competitive advantage it has over web and mobile. This book is also full of tips from fellow entrepreneurs, sharing their experiences while building a bot business.

How Is This Book Organized?

We'll start with an overview, then move on to the theory and deep dives into practical examples.

OVERVIEW

Chapters 1 through 4 provide an overview of the bot ecosystem—we define what bots are, what types of bots there are, which platforms they work on, and what use cases they can support. If you are already familiar with bots you might want to skim over these chapters.

THEORY

Chapters 5 through 13 review the different aspects that compose this new user experience and explain how to design each aspect. We start with thinking about the use cases and marketing, and move on to conversational elements that compose these experiences. We also talk about extended attributes such as distribution, engagement methods, and monetization.

PRACTICAL DESIGN

Chapters 14 through 19 are a step-by-step guide to designing bots. We will exercise the lessons learned and cover advanced topics like user testing and analytics. We will demonstrate designing a consumer bot as well as a business process bot.

TO INFINITY AND BEYOND

Chapter 20 takes a look at future trends of bot design, bot platforms, and market trends. Skip this chapter if you do not believe in predictions of the future.

O'Reilly Safari

Safari (formerly Safari Books Online) is a membership-based training and reference platform for enterprise, government, educators, and individuals.

Members have access to thousands of books, training videos, Learning Paths, interactive tutorials, and curated playlists from over 250 publishers, including O'Reilly Media, Harvard Business Review, Prentice Hall Professional, Addison-Wesley Professional, Microsoft Press, Sams, Que, Peachpit Press, Adobe, Focal Press, Cisco Press, John Wiley & Sons, Syngress, Morgan Kaufmann, IBM Redbooks, Packt, Adobe Press, FT Press, Apress, Manning, New Riders, McGraw-Hill, Jones & Bartlett, and Course Technology, among others.

For more information, please visit *http://oreilly.com/safari*.

Comments and Questions

Please address comments and questions concerning this book to the publisher:

> O'Reilly Media, Inc.
>
> 1005 Gravenstein Highway North
>
> Sebastopol, CA 95472
>
> (800) 998-9938 (in the United States or Canada)
>
> (707) 829-0515 (international or local)
>
> (707) 829-0104 (fax)

We have a web page for this book, where we list errata, examples, and any additional information. You can access this page at: *http://bit.ly/designing-bots*.

To comment or ask technical questions about this book, send email to *bookquestions@oreilly.com*.

For more information about our books, courses, conferences, and news, see our website at *http://www.oreilly.com*.

Find us on Facebook: *http://facebook.com/oreilly*

Follow us on Twitter: *http://twitter.com/oreillymedia*

Watch us on YouTube: *http://www.youtube.com/oreillymedia*

Acknowledgments

Writing a book was one of the hardest things I have done in my career. And I could not have done this without the support of my friends, family, and the amazing bot community.

So many people helped me with this book. Thank you to...

My loving wife, Deby Shevat, who had to learn about bots and conversational interfaces while reviewing my book (although it is far away from her profession, working with autistic kids), and my kids Daniel and Jonathan, who had to hear me talk on and on about the book, and kept supporting me while I was doing so.

My good friends and reviewers, Dana Cohen Baron, Chris Messina, Dr. Jacob Greenshpan, and Mike Brevoort, who provided me with a lot of feedback and insights.

All the bot entrepreneurs who contributed their experience to the book. You made it 100 times more valuable—thanks to Tomer Sharon, Greg Leuch, Nir Eyal, Dennis Mortensen, Rachel Law, Alyx Baldwin, Andy Mauro, Vittorio Banfi, Dennis Yang, Josh Barkin, Dr. Barbara Ondrisek, Oren Jacob, Dan Manian, Dmitry Dumik, Veronica Belmont, Mike Melanin, Artyom Keydunov, Peter Buchroithner, Ben Brown, Dan Reich, Mikhail Larionov, Lauren Kunze, Hillel Fuld, and Laura Newton.

All my colleagues at work who are building an awesome bot platform with me, and my manager April Underwood, who believed in me enough to let me do my day job while writing this book at night.

My amazing editor, Angela Rufino, and Mary Treseler, who believed I could write this book, as well as Rachel Head, my copyeditor.

And to you too, readers. I hope you find this book useful as you embark on your bot-building journey.

[1]

What Are Bots?

We come in peace.

—THE DAY THE EARTH STOOD STILL

BOTS ARE THE FUTURE OF SOFTWARE.

Bots are going to disrupt the software industry in the same way the web and mobile revolutions did. History has taught us that great opportunities arise in these revolutions: we've seen how successful companies like Uber, Airbnb, and Salesforce were created as a result of new technology, user experience, and distribution channels. At the end of this book I hope you will be better prepared to grab these opportunities and design a great product for this bot revolution.

Our lives have become full of bots in 2017—I wake up in the morning and *ask* Alexa (a voice bot by Amazon) to play my favorite bossa nova, Amy (an email bot by x.ai) *emails* me about today's meetings, and Slackbot (a bot powered by Slack) sends me a *notification* to remind me to buy airline tickets to NYC today. Bots are everywhere!

There is a lot of talk about bots these days, and a lot of misconceptions. In order to demystify these misconceptions, let's start by providing some of the history of bots and defining bots—what they do and why they are important.

I wrote my first bot 16 years ago. I was an engineer at a company that provided SMS infrastructure that was about to be deployed in Europe. You can imagine that testing if texting works in a network of one (as the system was not online yet, I was the only one on the network) is a very lonely experience. So I created a small program to answer my texts. It started as a bot that repeated everything I said—I would text "hello" and get a "hello" back—but that became boring really fast. I started adding a persona to the bot, adding funny sentences I heard in the office. At the end I had two personas I was chatting with constantly, "Bob" and "Samantha." I kept growing their vocabulary and skills and found it extremely therapeutic to converse with them via text.

But bots go way back to 1950s, when computer scientist Alan Turing contemplated the concept of computers communicating like humans. Turing developed the Turing Test to test a computer's ability to display intelligent behavior equivalent to that of a human. A user had to distinguish a conversation with a human from a conversation with a computer, and if they failed to do so, then the computer would have passed the Turing Test. Alan Turing was one of the fathers of computer science, and we still refer to the Turing Test when we talk about intelligent bots.

One of the best-known bots from the past was Eliza. Developed by Joseph Weizenbaum in 1964 for the IBM 7094, Eliza was a psychotherapist bot that talked to users about their problems, invoking strong emotional reactions in many users even though it was clear they were interacting with a bot and not a human.

So, What Are Bots?

At a very basic level, bots are a new user interface. This new user interface lets users interact with services and brands using their favorite messaging apps.

Bots are a new way to expose software services through a conversational interface. Bots are also referred to as chatbots, conversational agents, conversational interfaces, chat agents, and more. I will try to be consistent and call them bots in this book.

In many cases, bots are digital users within a popular messaging product, such as Slack, Facebook, Kik, and more. Unlike most users, they are powered by software rather than by a human, and they bring a product, a service, or a brand into a given messaging product via conversation. In this book we will focus on these cases, as they are the most common, but I want to acknowledge that there are other ways to expose a bot, and we will cover these when we talk about bot types in Chapter 2.

A common mistake is to think that the bot is the service itself—the bot is only an interface into the service, in the same way a website can be an interface for booking a flight. You can also use a mobile app to book a flight or call a human agent that can book that flight for you, all exposing the same service.

[KEY TAKEAWAY]

The bot is only an interface into the service, in the same way a website can be an interface for booking a flight.

Figures 1-1 through 1-3 show a few visual examples of bots and the services they expose.

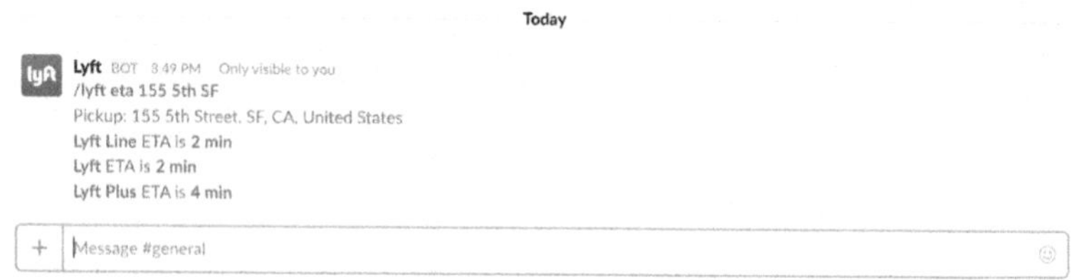

Pickup: 155 5th Street, SF, CA, United States
Lyft Line ETA is 2 min
Lyft ETA is 2 min
Lyft Plus ETA is 4 min
Message #general

FIGURE 1-1.
The Lyft bot providing a ride service time estimate to the user via the Slack chat medium

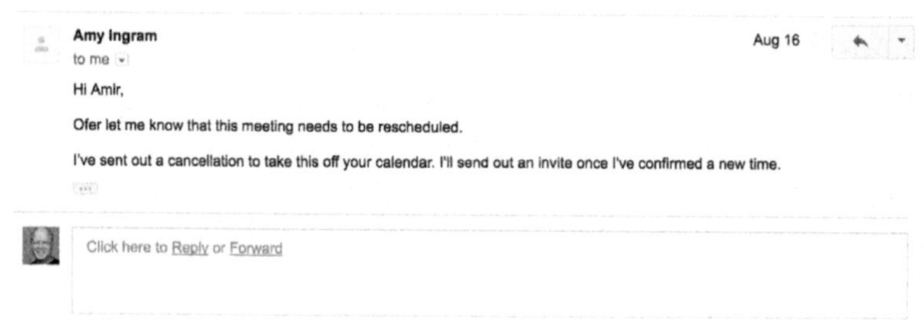

Amy Ingram
to me
Aug 16
Hi Amir,
Ofer let me know that this meeting needs to be rescheduled.
I've sent out a cancellation to take this off your calendar. I'll send out an invite once I've confirmed a new time.
Click here to Reply or Forward

FIGURE 1-2.
The Amy bot providing a scheduling service to the user via an email medium

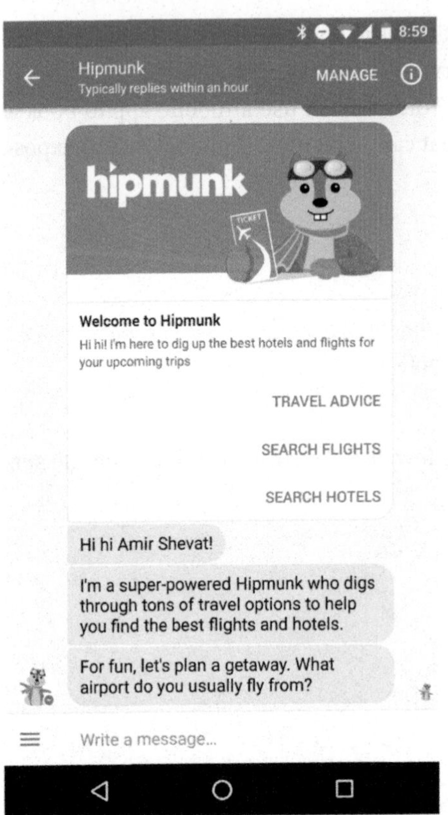

FIGURE 1-3. The Hipmunk bot providing a travel booking service to the user via the Facebook Messenger medium

Now that we have explored the origins of bots and seen a few examples, let's examine the market trend that led to this evolution.

The Bot Revolution and Evolution

Why do we need these bots? Why would we want to expose a service through a conversation? Why not just build a web page (or a mobile app), like we have been doing for the last 20 years or so? Isn't that much easier than building bots?

The answer is that things have changed in the software industry and in user behavior, and these changes are making bots more and more compelling to software companies. Here are some of the key developments we have seen:

1. In the last few years most users have adopted mobile, and it has become harder and more expensive to impress and engage with them though the web. This has made a lot of software providers

turn to creating native mobile apps (apps that run natively on your phone, for example Instagram or Google Maps) and exposing these mobile apps through app stores.

2. The mobile apps ecosystem quickly became saturated, making it harder and more expensive to compete. In addition, users became tired of installing and uninstalling mobile apps, and only a very few apps prevailed.
3. Surprisingly, the apps that prevailed and became very common were the messaging apps. Most modern users have three or more of these apps on their phone. (Open your phone and count the messaging apps you have there—don't forget to include SMS, email, and Siri/Google Now.)
4. User mindshare has stuck with the messaging apps. Users spend most of their time in these apps—this is even a growing trend with young users who do not have the "old" notion of the web, and spend most of their time in chat. Messaging and the ubiquity of connectivity mean that people are more available and responsive via messaging than alternative, indirect modes of communication.
5. These new apps opened up the ability to expose services, products, and brands on these chat platforms. Slack and Kik launched their platforms in 2015, followed by Facebook, Skype, and Apple in 2016.
6. In conjunction with these user and industry trends, technology has made a leap in natural language processing, making it easier (though not easy) to build and construct conversational interfaces.

Figure 1-4 sums up the interface's evolution.

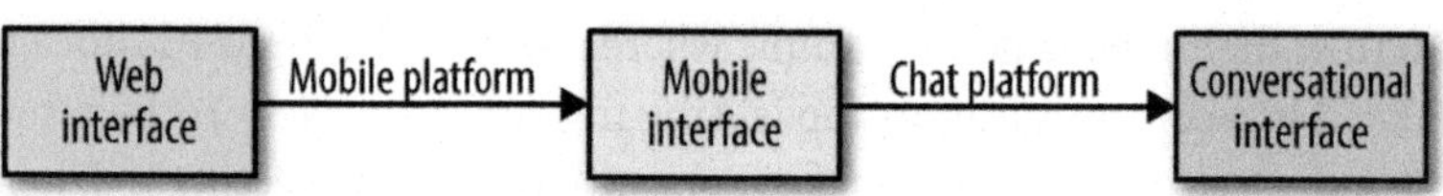

FIGURE 1-4.
From web to mobile to conversational interface

It is important to note that mobile interfaces were better than web interfaces in many ways and could facilitate new use cases (such as location-based services and camera-based services), but mobile was not better than the web for other use cases (such as long document creation).

The same thing is true for bots. As a designer you will need to explore which use cases are better for this new interface—bots are a great new hammer, but not everything's a nail.

Another way to look at bots is as *another* venue to engage with your users. You can still provide a dedicated website or a mobile app and integrate it with a messaging platform of your choice to re-engage with your user and expose different aspects of your service.

Figure 1-5 shows an example: an airline app exposing some of its services, such as buying a ticket, using a traditional web interface, and engaging with users later to report on flight time changes using the bot. The biggest value of the bot here is that users are already accustomed to getting notifications through their messaging apps, so the engagement rates are much higher.

[KEY TAKEAWAY]

The biggest value of the bot here is that users are already accustomed to getting notifications through their messaging apps, so the engagement rates are much higher.

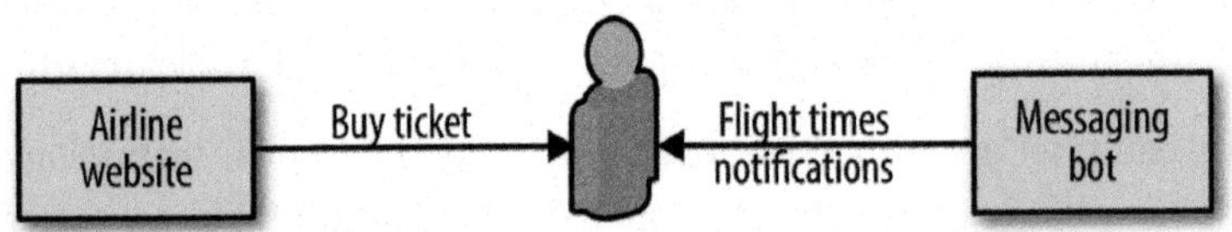

FIGURE 1-5.
Augmenting a service with a conversational interface

On the other hand, a good example where a service can be 100% encapsulated in a communication app is Amy Ingram, from New York–based startup x.ai. Amy is a personal assistant that sets up meetings. When people email me and want to meet, I add Amy to the email conversation (by CCing her email, *amy@x.ai*), and Amy does all the rest! The bot is connected to my calendar and knows when I am available; it has learned from me when and where I like to meet, both in person and online. Amy replies to about 300 emails a month on my behalf. We will talk about that use case at length in Chapter 6, and hear from the founders of x.ai to learn about their experience.

Just to show you the value of a bot like Amy, here's an example of a week's report (Figure 1-6).

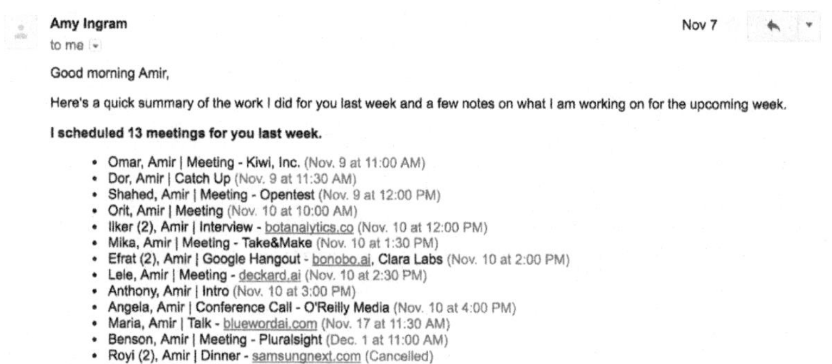

Amy Ingram
to me
Nov 7

Good morning Amir,

Here's a quick summary of the work I did for you last week and a few notes on what I am working on for the upcoming week.

I scheduled 13 meetings for you last week.

- **Omar, Amir | Meeting - Kiwi, Inc.** (Nov. 9 at 11:00 AM)
- **Dor, Amir | Catch Up** (Nov. 9 at 11:30 AM)
- **Shahed, Amir | Meeting - Opentest** (Nov. 9 at 12:00 PM)
- **Orit, Amir | Meeting** (Nov. 10 at 10:00 AM)
- **Ilker (2), Amir | Interview - botanalytics.co** (Nov. 10 at 12:00 PM)
- **Mika, Amir | Meeting - Take&Make** (Nov. 10 at 1:30 PM)
- **Efrat (2), Amir | Google Hangout - bonobo.ai, Clara Labs** (Nov. 10 at 2:00 PM)
- **Lele, Amir | Meeting - deckard.ai** (Nov. 10 at 2:30 PM)
- **Anthony, Amir | Intro** (Nov. 10 at 3:00 PM)
- **Angela, Amir | Conference Call - O'Reilly Media** (Nov. 10 at 4:00 PM)
- **Maria, Amir | Talk - bluewordai.com** (Nov. 17 at 11:30 AM)
- **Benson, Amir | Meeting - Pluralsight** (Dec. 1 at 11:00 AM)
- **Royi (2), Amir | Dinner - samsungnext.com** (Cancelled)

FIGURE 1-6.
Amy saved me time by coordinating meetings with 13 different people in a single week

Stages of Bot Adoption

Like with the web and mobile revolutions of the past, the software industry is transitioning through phases of adoption with the development of bots (Figure 1-7). The phases are:

1. *What are bots? Why do we need them?* At this stage most users and software providers are not aware of bots or how to use them (we are still in this stage, at the time I'm writing this book).
2. *We need a bot interface too!* At this stage a lot of software vendors start building bots. Most bots suck at the beginning, as there is very little experience in the industry. (We are rapidly moving toward this phase.)
3. *We are going with bot-first!* After a while, several bot builders experience success, and the bot user interface becomes common. Startups begin to adopt a bot-first mentality.
4. *We are bot only!* Some services are built with a bot-only user interface, and others with major parts of their workflows happening in conversation.

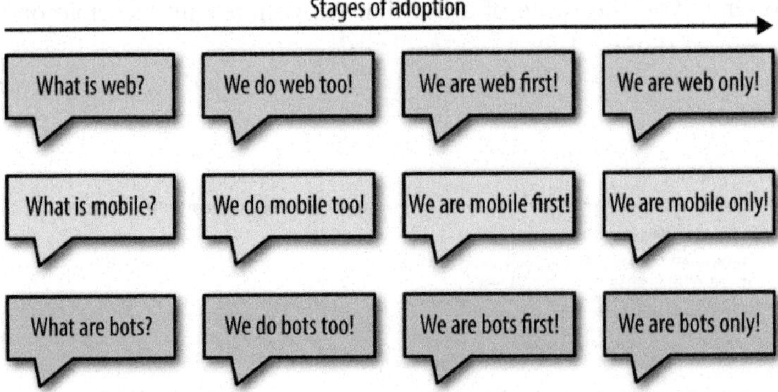

FIGURE 1-7.
The four phases of adoption

The same is true for user adoption. Until recently very few users (outside Silicon Valley) have been aware of bots, but that is beginning to change rapidly with Facebook, Kik, Google, Amazon, and other bot platforms driving consumer bots into the market and Slack promoting bots for work.

Not All Bots Are Born the Same

Bots differ from one another in many aspects. There are B2B bots and B2C bots, team bots and personal bots, and a huge amount of variation even within these spaces. Bots on Facebook are very different from the bots on Kik, even though they are both consumer bots. There are also *super bots* that expose a set of pluggable services rather than a single service, giving you the ability to add a skill to their bot rather than expose your own bot interface.

Closing Thoughts

We are at the dawn of a new technological era—an era in which software is going to engage with us on our own turf, and conversational interfaces are going to surface in more and more of the tools and services that we use every day. This revolution is going to change our work lives as well as our experiences as individuals and as a community.

In the next chapter we take a deep dive into different types of bots. This will give you a deeper understanding of the ways bots engage with humans and a better ability to decide which type of bot you want to design.

[2]

Bot Types

We are all unique, just like everyone else.

—**PROFESSOR DAN ARIELY**

As we discussed in the previous chapter, not all bots are the same. Let's examine the major types of bots out there. Understanding the different types of bots will provide you with the ability to pick the right type for your use case and allow you to explore alternative ways to expose your service, product, or brand.

Personal Versus Team Bots

A *personal bot* (also called a direct message bot or private bot) is a bot that is serving as a personal assistant. Communicating directly with the user, on a one-on-one basis, this bot has a single-user focus. An example would be a business travel bot that a user talks to directly in Slack, or a shopping bot in Facebook Messenger or Kik (Figure 2-1).

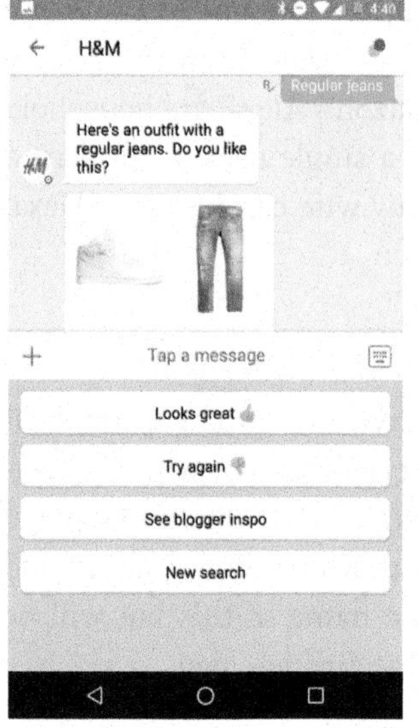

FIGURE 2-1. The H&M shopping bot on Kik

A *team bot* facilitates team processes and activities—for example, the Lunch Train bot (Figure 2-2) that helps teams choose where to go to lunch together, or the Standup Bot that facilitates team standups. A team bot can talk with multiple users either directly (privately) or publicly in a channel/group setting.

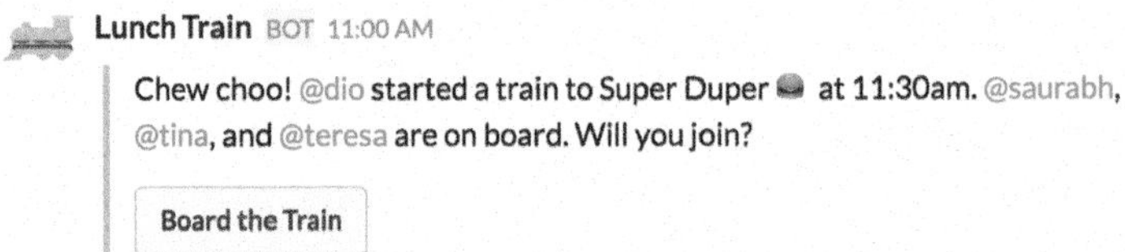
Lunch Train will send an interactive message to the channel or direct message
Lunch Train BOT 11:00 AM
Chew choo! @dio started a train to Super Duper at 11:30am. @saurabh, @tina, and @teresa are on board. Will you join?
Board the Train

FIGURE 2-2.
The Lunch Train bot—an example of a Slack bot that enables people to plan where to go to lunch together

Team bots are a little more complex to design: as multiple users can communicate with the same bot in a channel, the bot might need to juggle conversations. Personal bots might be easier to implement, but cover a more limited set of use cases. Personal bots could facilitate a process between multiple people, such as vacation/paid time off approvals, but they only communicate with a single user over a single context at a time.

Some platforms, such as Slack, provide you with the ability to take a hybrid approach and enable your bot to support both one-to-many and one-to-one engagement. Platforms such as Facebook currently only support personal bots. Some bots, like Amazon's Alexa, are household bots—they treat everyone in the house as a single user. I can start a song, my kids can raise the volume, and my wife can just say "Alexa, stop" and the bot will comply.

Super Bots Versus Domain-Specific Bots

A domain-specific bot exposes a single service (or product/brand). It represents that service, and the user associates the bot with that service. Let's imagine an airline travel bot (Figure 2-3)—it will help you with everything from booking flights to providing travel alerts and notifications on membership benefits. The name of this bot will be Airline-bot and the logo of the bot will be the airline's logo.

Amir Shevat 3:08 PM
hi
Airline-bot APP 3:08 PM
I am your Airline bot! I can help you book flights and get alerts for and

FIGURE 2-3.
This is how I would imagine my delightful Airline-bot

A super bot, on the other hand, is a single bot that facilitates and exposes multiple services. This bot may enable other services to plug in to it and extend its functionality.

Google Assistant is a great example of a super bot. It is a single bot that exposes different Google services, as well as data from Wikipedia and calls to action to Yelp and Google Maps. As you can see in Figure 2-4, Google Assistant provides a set of top services like maps, weather, news, games, and so forth, and subcategories of services inside these top services.

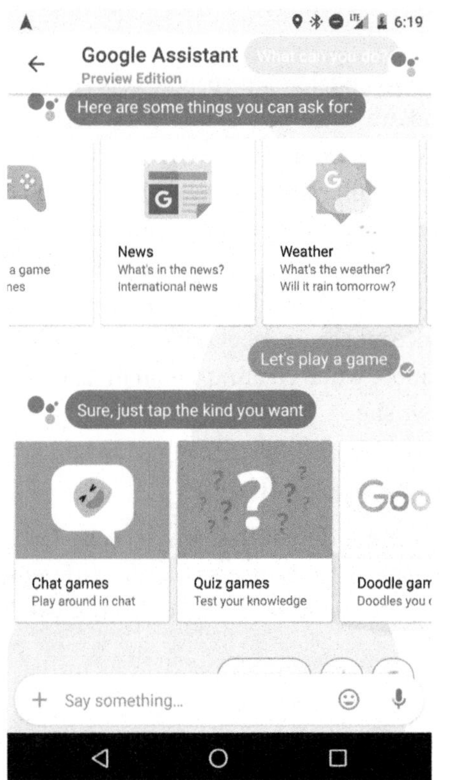

FIGURE 2-4.
Google Assistant in the Allo chat app

The advantage of a super bot is that the user does not need to deal with a lot of different bots, each addressing a different pain point or intent. Users don't need to remember the names of each bot, how to work with it, and how to access it—they have a single bot to work with, and that bot addresses all their needs. The super bot exposes all the services and provides the users with a consistent user experience.

The advantage of the domain-specific bot is that it can specialize for a specific type of content and intent. Users do not need to navigate through services and submenus to get to where they want to go to—the bot is there for one purpose and that purpose alone. It is also a more open architecture that lets software vendors expose their own brand and bot personality, as you can see with my example of the airline bot above.

[KEY TAKEAWAY]

The advantage of the domain-specific bot is that it can specialize for a specific content type and purpose.

From a design standpoint, super bots and domain-specific bots are very different. Super bots try to standardize the way services are exposed to users. When you want to integrate your service into a super bot (for example, as an Amazon Alexa skill), you need to understand that the super bot mandates the user experience—you as a designer have less control over how your service will be exposed. When you build a domain-specific bot, you have more control over the user experience. As an analogy, integrating your service into a super bot is like contributing an article to a third-party publication, while creating a domain-specific bot is like building your own news site.

Business Bots Versus Consumer Bots

Business bots and consumer bots are different in many aspects. They serve different purposes, they engage with the users in a very different way, and they even have different best practices around task length and outcome.

The purpose of bots for business is to facilitate a task or a business process in an easy, pleasant, and productive way. Communication should be to the point, with a focus on getting things done rather than talking about it.

Let's talk a little bit about task completion. Do users like to do expense reports? Do they flourish when they fill out vacation requests? If you've ever worked in an office, you know the answer is probably no. Bots for business try to remove these pain points by being everyone's own personal assistant. In the same way VPs in big companies have personal assistants who handle all the paperwork, bots strive to streamline these processes for the rest of the company's employees. You just send your receipts to the Expense-bot, and the bot takes care of all the rest—wouldn't that be magical?

The same goes for business processes. A bot can act as a program manager that coordinates between team members and facilitates complex business processes and workflows. It can interact with multiple workers to make sure things are created, approved, and shipped on time. Bots can assign bugs and follow up on the fixes all the way to production. They can follow up on a sales lead, all the way to completion and billing.

An example of a project management bot is one of the first business productivity bots, called Howdy (Figure 2-5). Howdy lets managers automatically collect information from their team members (replacing the standup meetings many managers hold daily) and brings that information back to the manager. Users can train the Howdy bot to run multiple question scripts.

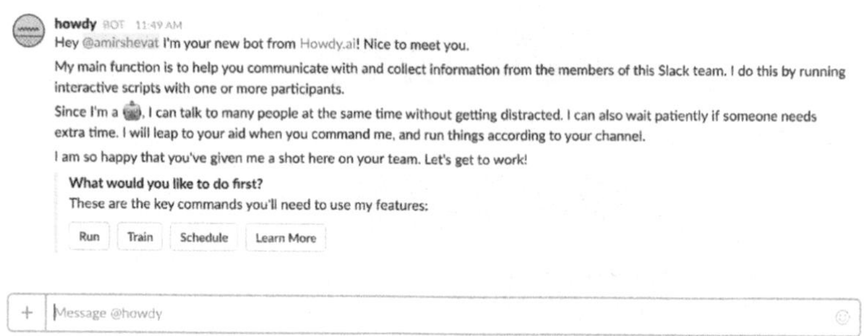
howdy BOT 11:49 AM
Hey @amirshevat I'm your new bot from Howdy.ai! Nice to meet you.
My main function is to help you communicate with and collect information from the members of this Slack team. I do this by running interactive scripts with one or more participants.
Since I'm a , I can talk to many people at the same time without getting distracted. I can also wait patiently if someone needs extra time. I will leap to your aid when you command me, and run things according to your channel.
I am so happy that you've given me a shot here on your team. Let's get to work!
What would you like to do first?
These are the key commands you'll need to use my features:
Run
Train
Schedule
Learn More
Message @howdy

FIGURE 2-5.
The Howdy team productivity bot

Consumer bots are a totally different story—their purposes are to entertain us, facilitate commerce, help us keep in touch with our favorite brands, stay up-to-date with news, keep in shape, improve our personal productivity and well-being, and more (all the other fun things we do outside of work). These are just a few examples of what consumer bots can do, and there are many more opportunities for bots to enrich and delight us in our daily lives.

Consumer bots can, in some cases, be chatty, whimsical, and personal—users tend to stray off topic more with consumer bots, and some bot builders actually measure the length of the conversation as a key metric for the success of the bot. Users are more tolerant of reengagement tactics like "what's new" notifications from consumer bots. In general, consumer bots are less task and workflow oriented and more experience oriented.

An example of an interesting consumer use case is bots that entertain you by chatting about any topic on your mind. This is a very popular use case amongst teens. Mitsuku (Figure 2-6) is a conversational bot on Kik that has won several awards around artificial intelligence and engagement.

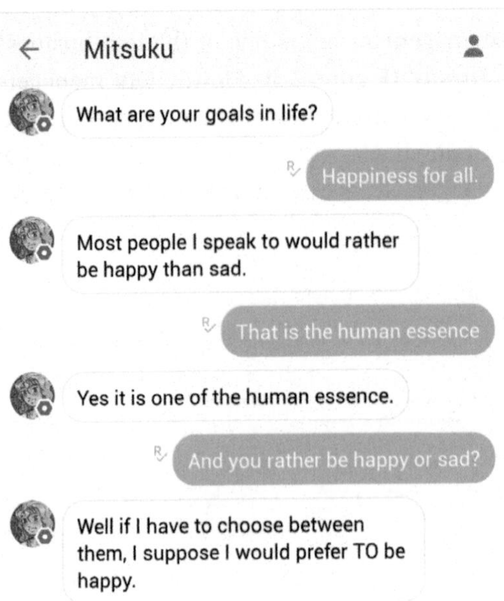

FIGURE 2-6.
Mitsuku talking about life—Mitsuku won the Loebner Prize for most humanlike AI in 2013 and 2016

There are a few consumer bots that need to be more similar to the business bots—joking with my bank bot about my finances is probably not a best practice. But in general, consumer bots need to be memorable while business bots need to be as transparent and minimal as possible.

[KEY TAKEAWAY]

Business bots and consumer bots are different in many aspects. They serve different purposes and engage with users in a very different way.

Voice Versus Text Bots

Another way bots differ is in the way you converse with them. Currently, bots support two major means of conversation: voice and text.

Text bots usually manifest themselves in chat apps. Slack, Facebook, Telegram, Kik, and WeChat are just a few examples of platforms where conversational bots are available. Some bot builders also build dedicated apps for their bot, an approach that might work for a few use cases but comes with a lot of app distribution challenges. Figure 2-7 shows an example of a text bot.

LunchBot BOT 5:01 PM
who is ready to order lunch? We have Pizza and Burgers today!

FIGURE 2-7.
LunchBot coordinates a team's lunch plans in a conversation

Voice bots include Amazon's Alexa, Microsoft's Cortana, Apple's Siri, Google Assistant, and a few more. There is usually a voice command or a button click that initiates the conversation, and the conversations are usually in the format of short commands or question/answer.

Amazon Echo (Figure 2-8) is a household device that exposes the Amazon voice bot. In my home, we all interact with this bot on a daily basis. Everything from "play the Beatles" to "set timer for 30 min" and "add milk to our shopping list" is done by voice interaction with Alexa.

FIGURE 2-8.
Amazon Echo exposing the Amazon Alexa voice bot

Although the use cases sometimes overlap, voice bots are usually great for hands-off experiences like driving, cooking, watching TV, and listening to music, while text bots are usually great for desktop and mobile engagement.

From a design perspective, voice bots are very different from text bots. If you want to take a deep dive into designing voice bots, I recommend reading *Designing Voice User Interfaces* (*http://bit.ly/designing-voice-user-interfaces*), written by my friend Cathy Pearl (O'Reilly).

Net New Bots Versus Integrations Exposing Legacy Systems

Another distinction to notice is the difference between integration bots and new services exposed as bots.

The underlying assumption with integration bots is that we have a core service that is not going away anytime soon, that we need to expose in a conversational interface to improve engagement, usability, and brand recognition. The bot is usually branded with the legacy system's branding, and the major design challenge is *which part of the functionality should we expose in this bot?* Integration bots usually start with very light functionality, which grows as the bot becomes more successful.

Figure 2-9 shows an example of a very common integration between Slack and a customer relationship management (CRM) bot.

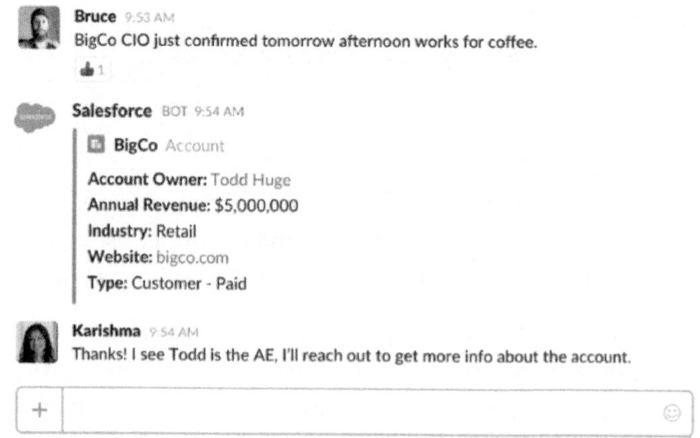

FIGURE 2-9.
The Salesforce bot adding customer information to a conversation in Slack

As you can see, the CRM bot pulls relevant information from Salesforce that is specific to the context and posts it in the conversation. As you might guess, there is a lot more information in the CRM system for that client, but the bot only posts a concise and contextually relevant snippet of the available information.

Net-new bots are the primary interface of a new service or product. The major design challenge here is *how can we expose all of the functionality we need in a conversational interface?* From my experience, like with many new services, the minimum viable product (MVP) approach applies to these new bots—start with single focused value proposition and grow from there.

An example of a net-new bot is Growbot (Figure 2-10). The Growbot team had an idea about how to facilitate team culture in new way. This is a company that built its business—an HR bot—bot-first, providing a conversational interface from day one.

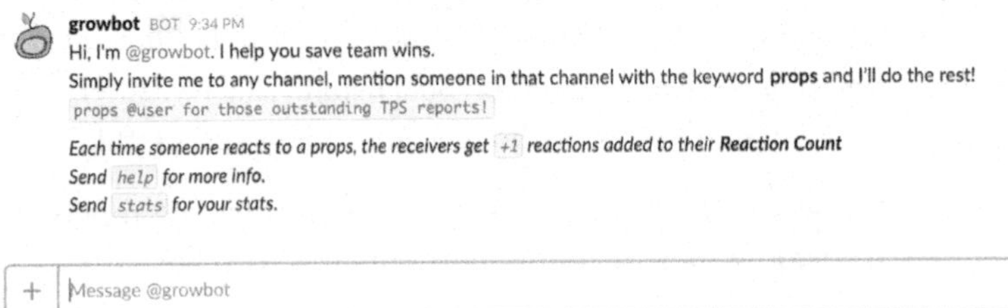

growbot BOT 9:34 PM
Hi, I'm @growbot. I help you save team wins.
Simply invite me to any channel, mention someone in that channel with the keyword props and I'll do the rest!
props @user for those outstanding TPS reports!
Each time someone reacts to a props, the receivers get +1 reactions added to their Reaction Count
Send help for more info.
Send stats for your stats.
+
Message @growbot

FIGURE 2-10.
Building team culture—Growbot is a bot-first company

Closing Thoughts

Bots are a new user interface that can be implemented in many ways and solve different types of business and consumer use cases. They can extend and augment existing services and expose new ones.

In the next chapter, we will go over the major bot platforms and review the feature set and capabilities each platform provides to give you a better understanding of how these bots behave in their own habitat.

We will also suggest a process for picking the right platform to start with, by analyzing different aspects of your service or product and comparing them to the platforms' capabilities.

[3]

Major Platforms

Without habitat, there is no wildlife. It's that simple.
—**WILDLIFE HABITAT CANADA**

IN THIS CHAPTER WE'LL explore bots in their habitats. As a designer, you need to choose the right platform for your bot. There are many platforms that host bots, and we cannot cover all of them. We will pick a representative platform for each unique user experience and review the key aspects of each of these platforms. We will give a brief description of the bot capabilities in these platforms and cover these UI capabilities in more detail later, in Chapter 9. In the second part of this chapter we will discuss practical ways to pick the right platform(s).

The Business Bot Platform: Slack

A popular messaging platform for teams at work, Slack is available on mobile and desktop, serving tens of thousands of businesses from small startups to large enterprises. Slack users are very engaged, having Slack open about 10 hours per day on average. Business users pay for Slack, making the audience for your bot qualified and highly engaged. (Full disclosure: I work at Slack.)

The Slack API provides a wide range of actions that bots can do on the platform. These include:

- Post messages. Bots can send messages into Slack either publicly, to a channel, or privately, to a person (direct message, or DM) or set of people (multi-party direct message, or MPDM). Bots can post content that includes rich text, emojis, images, and more.
- Receive user and team message inputs, both text and files, in a specific channel, DM, or MPDM.

- Expose slash commands. A slash command is a unique command, following the pattern of */<command-name> <arguments>*, that invokes a response from the bot. An example might be a */report sales* command that will make the bot respond with a sales report.
- Expose buttons. These are clickable controls inside messages that can invoke actions on the service side.
- Subscribe to the Events API in Slack. Bots can be notified about events, such as when a user is added to a channel, leaves a channel, replies to a message, and so on.
- Use Slack as an identity provider by signing in with a Slack account.
- Perform administrative actions. Bots can provision channels, invite members, edit and delete messages, and more on behalf of the installing user.

The Consumer Bot Platform: Facebook Messenger

With more than a billion users, Facebook Messenger is a leading consumer platform for bots. Facebook bots interact with the user via the Messenger interface (Whatsapp, Instagram, and interaction within the Facebook main feed are not currently supported, but ads on the main Facebook feed can lead to bot interaction). Facebook Messenger is available on mobile and desktop.

The Messenger API provides the following rich functionality:

- Posting content—support for text, images, files, and structured templates that provide a consistent experience across bots
- Delivered callbacks—a bot can detect that a user has received messages
- Receiving content—a bot can access messages that the user inputs in the chat with the bot
- A rich set of predefined button actions, including Buy, Share, Call, URL, and Postback (to send an action to your bot)
- Quick Replies that provide the user canned responses to questions
- Opening a webview for custom out-of-Messenger interaction
- Sending geolocation information with a single click

The Voice Bot Platform: Alexa

Alexa is a super bot by Amazon that exposes multiple products and services. Most commonly, Alexa can be found in a device called Echo. Echo sits in your home and waits for you to call upon Alexa. Interactions with Alexa are vocal and usually sound like this: "Alexa, weather in San Francisco." Out of the box Alexa supports a variety of built-in capabilities such as music library, timer, weather, and search services. Alexa also supports third-party integrations through *Alexa Skills*, an API that provides developers with the ability to add voice commands to it.

The Alexa Skills Kit provides the following functionality:

- Register a secondary voice command called an *invocation name*. For example, your print service can register "Alexa, print *xyz*."
- Receive client inputs. The service transcribes the user's voice and sends it to you.
- Output voice back to the user. Alexa will read out the text you reply to the user with.
- Support for smart home skills—i.e., integration with Internet of Things (IoT) devices such as smart lights and connected home locks.

Note that this is not a platform that lets you add your own bot. You will need to plug in your service as a skill that the super bot exposes. Also note that Alexa is becoming available in other devices, such as smartwatches and third-party IoT devices.

The Teens' Bot Platform: Kik

Kik is a mobile messaging platform with more than 300 million users, targeting youths who like to chat and share content with their friends. Kik's emphasis is on brand engagement, letting teens engage with and follow their favorite brands.

The Kik API provides the following rich functionality:

- Sending messages, including text, links, images, and rich media
- Received/read and delivery receipts—the bot can detect that a message has reached the user's device and that the user has read the message

- Receiving messages—the bot can receive text messages posted by the user in a direct communication or messages that include an @ mention of the bot name
- Canned responses in the form of buttons
- Broadcasting a message to a large number of users in a low-priority, outside-of-the-conversation context
- Opening a webview for custom out-of-Kik interaction

The Legacy Bot Platforms

There are also a few more traditional bot platforms that we should consider. While you may not immediately think of these as bot platforms, they are actually very common and quite effective platforms for bots.

EMAIL

Email is a very common and standard means of communication. Many businesses use email as their sole communication platform. Emails are also common in business-to-consumer communication: from Zendesk support to MailChimp marketing engagement, businesses commonly use emails to interact with their partners and clients.

Both common email protocols, IMAP and POP3, provide a limited set of functionality:

- Sending emails to a user or a set of users (hiding some recipients using the BCC feature). Bots can email rich content that includes rich text, titles, emojis, images, and more.
- Received/read receipts. Using a hidden tracking pixel, bots can get notified when a user opens an email. This is not a 100% effective solution, as some clients block that pixel.
- Receiving emails (both new messages and replies to email threads). The bots can also reply to received emails.

SMS

The most common communication apps in mobile, SMS (Short Message Service, sometimes just referred to as text) apps use the cellular infrastructure rather than the internet, making them accessible and extremely popular in emerging countries and on low-end phones around the world. SMS services are tied to your phone number, making it somewhat easier to register with bots that use SMS as a medium.

The SMS API provides the following limited functionality:

- Sending short text messages (length depends on language encoding)
- Receiving short text messages (length depends on language encoding)

Some providers also support rich interactions such as sending and receiving images though the MMS (Multimedia Messaging Service) protocol, but that is usually unreliable and operator dependent.

How to Choose a Platform

Picking the right platform is critical to the success of your bot. A bot in the wrong habitat will shrivel and die. I see a lot of examples of bots that get very little engagement, low installation figures, and lots of complaints from users, just because they have a different state of mind than the developers intended.

[KEY TAKEAWAY]

Picking the right platform is critical to the success of your bot. A bot in the wrong habitat will shrivel and die.

Choosing the right platform is hard because the decision is based not only technical, business, marketing, or design considerations, but rather a combination of them all. Note that you can choose to offer your bot on more than one platform, although I recommend starting with one platform and then moving to the next.

Let's take two examples and go through the process of choosing a platform for each:

Gamez-bot

Gamez Inc. is a company that provides popular casual games like trivia, quiz, and turn-by-turn games.

Timez-bot

Timez Inc. is a time and attendance tracking and timesheet service provider.

EXPLORATION STAGE

Here are a few criteria that can lead you to the right decision.

Audience

First, define your audience and use case. Are you addressing a business use case? Is this a consumer use case? Are you targeting teens? Families? Adults at work? When are they using the service?

Gamez-bot

Gamez Inc.'s core customers are women aged 25–45, playing mainly in the evenings.

Timez-bot

Timez Inc.'s core users are tech-savvy mobile and distributed workforces, men and women aged 21–55, mainly using the apps from 7–10 a.m. and 6–8 p.m.

Consumer bot or business bot

Now, try to figure out if you are addressing a business use case or a consumer use case. Sometimes this is a very easy question to answer—for example, for a movie bot—but in some cases the answer is not so clear, like in the case of a travel bot. Also ask yourself how this bot will make money. Via subscriptions, affiliations, ads, in-bot payments? I would say that even in the case of a travel bot, building a business travel bot would be very different than a consumer travel bot, and this focus is super important. The answer "both B2B and B2C" is probably the wrong one.

Gamez-bot

Pure consumer play, making money by selling power-ups in the games.

Timez-bot

Small to medium business solution with a yearly license fee per seat.

Feature availability

Then, look at the conversational controls you will need to successfully deliver your service. Will voice interaction be sufficient for this task? Or do you need to visualize content? Do you need to provide fixed choices and actions, in the form of buttons or canned responses?

Gamez-bot

All games are basically text based, but management would love to use a carousel to promote different games and to improve sharing.

Timez-bot

Requires buttons to set and approve time/attendance as well as the ability to display reports, notify users who have not filled in their time/attendance reports, and set up team reminders and reports.

Preferred devices

Lastly, explore your engagement channels. Are you interested in accessing the users across devices? Is this a mobile-only service? An at-home service, or an on-the-go service?

Gamez-bot

Most users access the games via mobile; a small percentage use the desktop.

Timez-bot

Most users access the service via the web; a small percentage use the mobile app.

EVALUATION STAGE

Now go back to each of the platforms described above, and compare their capabilities, audiences, and engagement channels with your service requirements. In Chapter 9 we will provide a more in-depth review of each UI event.

Some of your criteria should be hard (like whether this is a B2B or a B2C bot) and some of your criteria can be soft (for example, the ability to use/display bold text). Try to make sure your platform of choice hits all of the hard criteria and as many soft criteria as possible. Sometimes it is important to get feedback from other stakeholders, such as marketing and engineering.

Also check that your service adheres to the specific platform's terms and conditions. There have been a few instances of bots that were rejected in the platform review process, because the services they exposed were not allowed by the terms and conditions of the platform they targeted (for example, serving ads is not allowed in some platforms).

Gamez-bot

It looks like Gamez Inc.'s core audience can easily consume the games on Facebook. The functionality fits the requirements. At a later stage, consider expanding to Kik in order to reach a younger audience.

Timez-bot

The audience is most reachable on Slack. The functionality fits the requirements. In cases where users are not on Slack, the bot will default to text/email.

VALIDATION STAGE

This is super important—validate your decision by talking to potential users. Ask them which tools they use to complete tasks or access services like the one you are planning to launch. Try to create a prototype (you'll see how in Chapter 17) and have them access the service and converse with your bot. I cannot stress enough the importance of the validation stage and the impact of it on your bot design and platform choice.

[KEY TAKEAWAY]

Validate your decision by talking to potential users. Try to create a prototype (you'll see how in Chapter 17) and have them access the service and use your bot.

Finding the right users is not easy—picking your friends and family will generally yield inaccurate feedback. Try defining your audience and then seeking people who meet those criteria. Google sometimes does that by putting its researchers on the street, but you can also use social media polls and even research companies to do this validation for you.

Gamez-bot

Target audience reacts very well to early validation. Some fine-tuning is needed in the trivia game; consider the use of canned responses to reduce confusion.

Platform chosen: *Phase 1*: Facebook Messenger. *Phase 2*: Explore Kik.

Timez-bot

Users love the new solution, which is much better than the Timez app (which crashes on iOS all the time). Some feedback on formatting and wording of the conversation.

Platform chosen: *Phase 1*: Slack. *Fallback*: text/email.

Now that you have validated your assumptions with potential users, you can start designing your bot to suit the specific platform you have chosen.

Closing Thoughts

Choosing the right platform to run your bot on is a critical factor to the success of your bot and your business. While you can launch on multiple platforms, choosing which platform you launch on first should be a thoughtful and well-validated decision. The steps we have described here are not much different from those involved in vetting any other app or web idea, so most product managers should be accustomed to these processes.

Before we dive into the design of a conversational interface, let's look at major use cases in the market today. Although we are still in the experimental stage of bots, there are interesting emerging use cases you can learn from.

[4]

Major Use Cases

The hard question is not "Are we doing it right?"; it is rather "Are we doing the right thing?"

—TOMER SHARON, WEWORK

LET'S EXPLORE SOME OF the use cases where bots can be utilized. This is not an exhaustive list by any means, but in order to start thinking about design aspects of your bot you will need to get a taste of what is out there.

Conversational Commerce

From buying on Amazon to ordering a ride, conversational bots can facilitate commerce in our lives. When done right, conversational commerce can be more intuitive and engaging than traditional commerce. No more shopping lists on the fridge—you just say "Alexa, add sugar to the shopping list." Travel bots can replace travel apps and websites, providing everything from booking to alerts of flight times, and customer service as well.

Users no longer need to install an app to get a ride—they can just ask their favorite @Uber or @Lyft bots for a ride in their chat app. This means that, for the first time since the mobile revolution, there is a clear separation between intent and installing an app. Assuming discovery is done right, this will also mean that the user acquisition cost for commerce will go down, possibly providing a more cost-effective means to reach your users.

A very interesting bot in this space is Kip (Figure 4-1), a shopping bot for teams. From office supplies to snacks, Kip handles the complex coordination of getting everyone in the team to add to the group order.

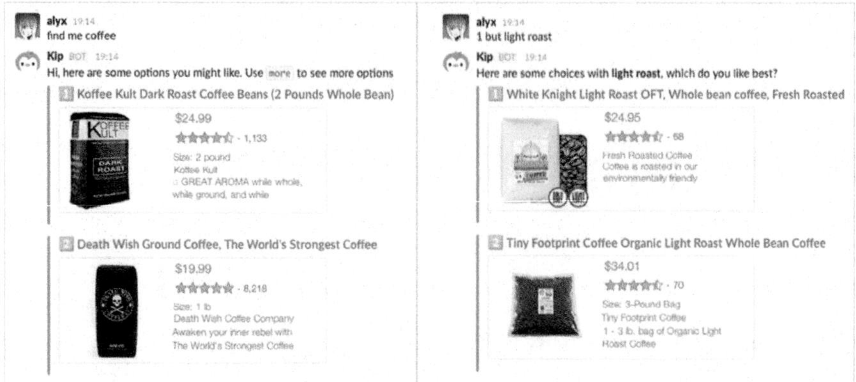

FIGURE 4-1.
Kip shopping bot on Slack

The interesting thing in this example is that the bot is introducing a new ecommerce concept called the "team cart," in which multiple members in the chat can add to the cart and the admin can pay for it. In this way Kip is enabling a way of shopping that was not available until now.

Bots for Business

This is where chat platforms for work, such as Slack, focus their efforts. Here we can find bots for HR, legal, sales, marketing, facilities, product, engineering, and other departments. GitHub has coined a term for its way of managing DevOps through chat: *chatOps*. Most startups connect their customer relationship management (CRM) systems to get notifications for new clients. Entire business operations can become more productive with personal assistants that help us do our work better.

Figure 4-2 shows an interesting legal bot use case.

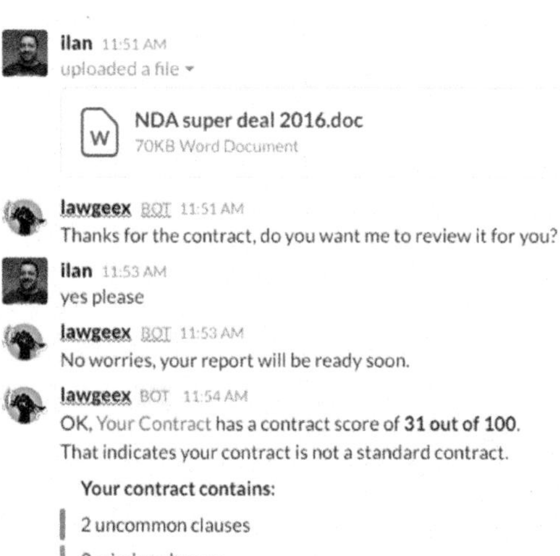

ilan 11:51 AM
uploaded a file
NDA super deal 2016.doc
70KB Word Document
lawgeex BOT 11:51 AM
Thanks for the contract, do you want me to review it for you?
ilan 11:53 AM
yes please
lawgeex BOT 11:53 AM
No worries, your report will be ready soon.
lawgeex BOT 11:54 AM
OK, Your Contract has a contract score of 31 out of 100.
That indicates your contract is not a standard contract.
Your contract contains:
2 uncommon clauses
3 missing clauses
17 important clauses
View full report

FIGURE 4-2.
The LawGeex legal bot

Here, the legal bot LawGeex is reviewing an NDA (non-disclosure agreement) contract and providing feedback to the legal team.

There is a strong incentive to use bots for business workflows. Most corporate workflows are cumbersome, require logging into legacy systems, and are often time-consuming. Using bots to facilitate short, contextual, and actionable tasks can greatly improve the productivity of a team.

[KEY TAKEAWAY]

There is a strong incentive to use bots for business workflows—facilitating short, contextual, and actionable tasks can greatly improve the productivity of a team.

Productivity and Coaching

Personal and professional productivity is a growing market. Here we can find bots focused on reminders and to-do lists, personal and team task management and completion. While these seem like simple use cases, they are very popular in the mobile app world and appear to have high engagement and install rates in the bot stores.

I also see many use cases for personal bot coaches that help users with weight loss, finances, parenting, sports, and more. It seems like the nature of the medium—having the bot talk to you in your chat app—makes the interaction more effective and engaging.

A good example of a coach bot is Lark (Figure 4-3).

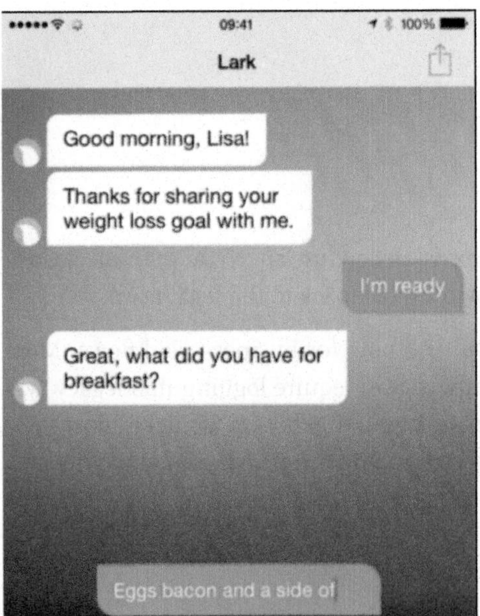

FIGURE 4-3. Lark helps users monitor their food habits and keeps them on track on the path to weight loss

One of the incentives for using bots in these types of use cases is user compliance—bots can provide a more personal experience that is harder to ignore, compared to mobile apps for example, and users are often more willing to provide information to a bot than to fill forms in an app.

Alert/Notification Bots

Bots for this set of use cases are starting to replace email or in-app notifications. These could be news bots, price watch bots, analytics report bots, or bots that notify you when your kids get home. There are a couple of differences between bot notifications and traditional mobile notifications:

- Notifications sent to a group/team/channel chat are more collaborative, and we see teams collaborating and taking action faster and more productively than when emails are sent to a group.
- While traditional notifications take you back to an app or a website to take action, many chat platforms provide you with a set of controls, such as buttons, that you can use to take action inline.

Given the right use cases, notification can very quickly turn into taking productive action (Figure 4-4).

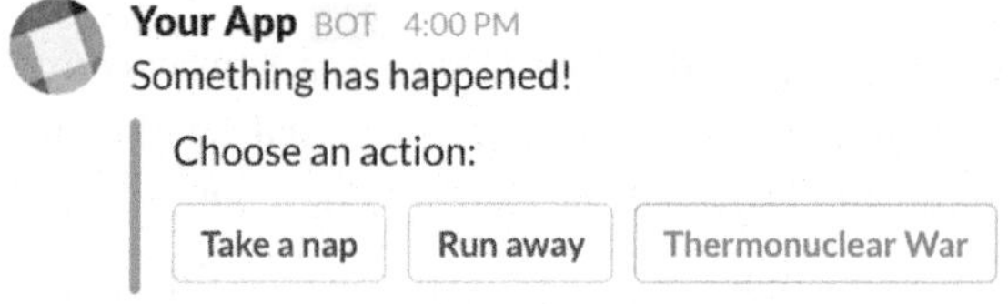

FIGURE 4-4.
Conversation can also lead to action

These micro workflows can happen in consumer use cases, such as a discount alert with an action button to buy, as well as business use cases, such as for actionable reports or approval processes.

The incentive here is that using bots for reports and alerts improves the actionability, transparency, and context. One alert in the #DevOps channel is worth a thousand emails.

Bots as Routers Between Humans

This is an interesting set of use cases where the service is actually provided by humans, but the bot acts as a router/operator and connects the user with the human service provider. In the same way that Lyft and

Uber connect you to a human driver (at least, at the time this book is being written), a bot can connect you to another human who then facilitates anything from IT support to songwriting.

A good example of an operator bot is Sensay. The Sensay bot (Figure 4-5) lets you instantly connect with a real human whenever you need advice or inspiration. It works across platforms and across devices.

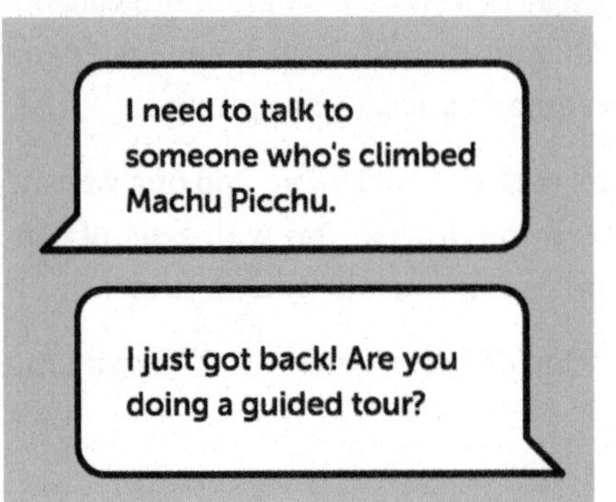

FIGURE 4-5.
An intent captured by Sensay—the bot connects the user to a human assistant

While some of the services provided by the humans Sensay connects users to would be hard to replicate with a bot, the actual act of connecting people is mundane and can easily be executed by a bot.

The incentive here is to provide a more friendly and useful version of the common interactive voice response (IVR) systems we all love when calling our service providers. The hope is that text-based bots that are delightful, personalized, and actually get us to the right person can change our negative perception of most common answering machine–like IVR systems.

Customer Service and FAQ Bots

This is one of the most common use cases for bots. Here, the bot serves as the first line of support, for internal employees or external customers. For internal use cases, the bot can answer questions like "What is our vacation policy?" An external consumer brand bot can answer questions like "What are your business opening hours?" Support and FAQs are an easy use case because they usually follow a pattern of a single request/response, and the questions are usually repeated and easily trainable.

There is a strong incentive to use bots in customer support use cases—this is because bots are typically much more cost-effective (and in many cases faster) than humans at performing simple repetitive tasks. These business use cases are very popular for bots on Facebook Messenger and Slack. According to initial experiments, bots can easily cover approximately 40% of internal and external support tickets.

Third-Party Integration Bots

At the time of writing of this book, the most requested bot by Slack users is one that integrates Salesforce CRM with Slack. Business users keep telling us that they want to be able to run account lookups from within Slack while talking about clients.

CRM is by no means the only system integration requested, though. From Google Analytics to Merkato, WorkDay to Concur, users crave simple integrations that will save them time and make them more productive.

Statsbot (Figure 4-6) is a great example of an integration bot. It pulls information from Google Analytics, Mixpanel, and other marketing systems and integrates the insights from these systems into Slack.

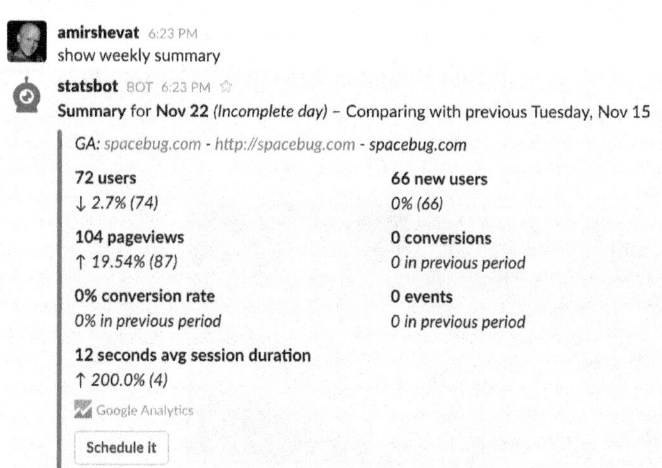

amirshevat 6:23 PM
show weekly summary
statsbot BOT 6:23 PM
Summary for Nov 22 (Incomplete day) – Comparing with previous Tuesday, Nov 15
GA: spacebug.com - http://spacebug.com - spacebug.com
72 users
↓ 2.7% (74)
66 new users
0% (66)
104 pageviews
↑ 19.54% (87)
0 conversions
0 in previous period
0% conversion rate
0% in previous period
0 events
0 in previous period
12 seconds avg session duration
↑ 200.0% (4)
Google Analytics
Schedule it

FIGURE 4-6.
Statsbot pulling information from third-party integrated systems

The core incentive here is that users do not want to context-switch between apps to get the information they need or run their workflows. They want to converse with the tools and services they use for work inside their chat apps.

Games and Entertainment Bots

These use cases are exploring ways to entertain and delight the user. Entertainment is a big part of our lives, and bots can be a part of that—from full-featured games on Kik to Alexa telling my kids fun facts and knock-knock jokes.

These bots are taking a different path from our traditional concepts of entertainment: neither very passive, like a TV, nor very rich and engaging, like a game console. The bots are trying to turn a conversation into a fun activity—go figure! Who would have thought conversation could be fun?

> **Fun fact:** Early experiments done with kids' movies showed that users were able to have long conversations with bots about their favorite characters and movies—sometimes even longer than the movies themselves.

An example of a social entertainment bot is the Swelly bot (Figure 4-7). Swelly lets you pick between two options and shares the voting results of all users. It is a delightful experience to casually vote on foods, fashion, vacation spots, and more.

FIGURE 4-7.
The Facebook Messenger Swelly bot asks, do you like pizza or lasagna?

Swelly also has a (non-bot) mobile app for both Android and iOS, but the team reports strong engagement on the bot user interface. Bots can reengage with users in an easy way and drive them back to the conversation or game. One of the core incentives here is that bots can reengage with the users and encourage them back to the service in a less intrusive and more customizable and friendly way than app notifications, for example.

Brand Bots

In this set of use cases, bots try to use the chat medium to create brand awareness and engagement. As bots become more popular and gaining traction with apps becomes more and more expensive and difficult, marketing managers are seeking ways to build bots for their brands.

There are some interesting use cases around notifications of new products or discounts by top brands, and a lot of other experiments. Bot builders are still trying to figure out what a valuable and engaging brand bot looks like over this new conversation interface.

Remember that bots are only as good as the services they expose, and bots for brands are no different. The Whole Foods Market bot in Figure 4-8 is a good example of a bot that not only provides access to the brand, but also adds value for the user.

[KEY TAKEAWAY]

Remember that bots are only as good as the services they expose, and bots for brands are no different.

The core incentive here is app fatigue—users are tired of installing specific brand apps. Bots provide brands a new and fresh way to engage with their users in a useful way.

Breaking Down Bots

As mentioned, like with most technologies and user interfaces, there are several components we will need to design, aspects we will need to consider, and decisions we will need to make as part of building our bots.

We will cover each of these aspects in depth in this section of the book, but first we'll start with a high-level view of bot anatomy (Figure 5-1).

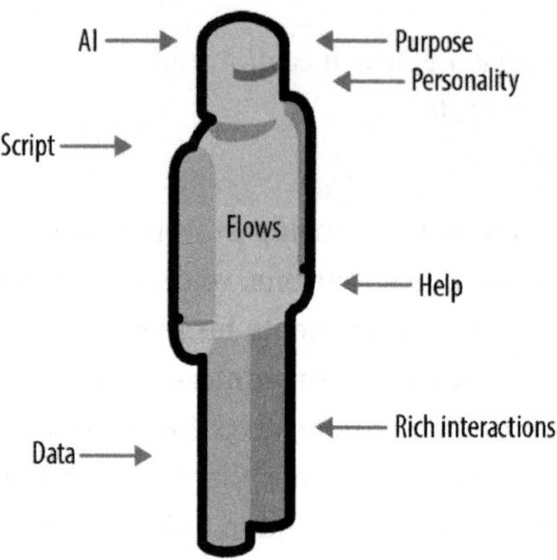

FIGURE 5-1.
Bot anatomy at 10,000 feet

The following attributes will explore different aspects of the bot's anatomy:

1. Branding, personality, and human involvement. Features include:

    a. *Personality*—Before starting to script, you need to decide what type of personality you want to bestow on your bot. This should be suited to the type of audience you want to address, the type of task you need to complete (getting things done versus having fun, for example), and the brand you want to associate this bot with.

    b. *Logos and icons*—As bots are a transparent(ish) UI, having a logo and an icon allows the user to identify the bot, which contributes to brand recognition. The bot's logo can also imply gender, age, and other human-like attributes.

c. *Naming*—The name can be as easy as a simple association with your brand. In other cases, naming a bot with a human name can create a stronger emotional connection. Naming can also have the same complexity and implications as logos when it comes to gender, age, and other attributes.

d. *Human intervention*—Routing a conversation to a human is quite easy, and can be transparent to the user in chat conversations. In some cases, having a human review bot answers, suggest course corrections, and handle errors might be a good initial strategy to manage the conversations, at least until your bot can manage these tasks without any human intervention.

2. Artificial intelligence (AI). Depending on the use case and type of conversation, artificial intelligence can be key to the success of your service by facilitating natural language understanding, conversation optimization, and many other aspects of your bot interactions. Elements to consider include:

   a. *Natural language understanding*—Understanding intents and extracting key variables (entities) from a user's inputs.

   b. *Conversation management*—Managing complex, multi-intent conversations.

   c. *Image recognition*—The ability to recognize text, objects, and even people's emotions in photos.

   d. *Prediction*—The ability to predict the right answer to a question, or an action to take at a particular time in the conversation.

   e. *Sentiment analysis*—The ability to understand the sentiment of the conversation.

3. The conversation. There are different aspects of the conversation to consider:

   a. *Onboarding*—A crucial part of the bot's success. Here you relay information to the users about the bot's purpose, ways to interact with the bot, what functionality is provided by the bot, and how to get help.

b. *Functionality scripting*—This is the meat of things. Here you script the *flows* (sometimes called stories) for each function, including happy paths and mitigations for failure. This is where you dive into the different types of conversations and talk about design best practices.

c. *Feedback and error handling*—This is an important part of the script that is sometimes overlooked. Feedback is one of the keys to making your bot better, and appropriate handling of failures is key to a good user experience as well as a way to improve your bot as time passes.

d. *Help and support*—At any time during the conversation, the user might get lost or thrown out of the happy path or flow (the main expected flow). Providing support and help can ensure a smoother usage of your bot.

4. Rich interactions. The bot may need to support any of the following:

   a. *Files*—Both bots and users can upload files to the conversation, in most messaging platforms. Examples might be work documents or shopping lists.

   b. *Audio and video*—Rich media can be the core functionality of some bots, and can enrich conversations in other use cases.

   c. *Images, maps, and charts*—As conversations are not super-rich environments, images can enrich the experience and entice the user to take action, as well as provide a lot of information that it would be very hard to relay with text.

   d. *Buttons*—These controls take the form of canned responses in some platforms and full controls in others. In any case, buttons can help users complete tasks faster by circumventing lengthy conversations.

   e. *Templates*—Some platforms provide a set of more complex and rich templates, such as message attachments and carousels. These templates help with standardization of common elements and make the user experience more predictable.

f. *Links and formatting*—On the same note of making the conversation more engaging, formatting and adding links can improve engagement and retention. Formatting a message, from color coding to font styling, can relay intent, convey state and progress, and direct the user to the right path.

g. *Emojis and reactions*—Emojis are a great way for the bot to convey information about states such as task completion or failure and to relay emotions. In some platforms the bot can add emoji reactions, send emojis, and use emojis in the text.

h. *Persistent menus*—As the user may get lost in the conversation, cancel a conversation, or context-switch to another task, you will need to think about giving your users a solid understanding of how to navigate the bot conversation.

i. *Typing indications*—This feature enables the bot to fake typing events, giving the user the impression that the bot is typing a response. This helps give the user a sense of the bot's presence.

j. *Slash commands*—These are easy shortcuts to invoke actions in a command line–like manner.

k. *Webviews*—This feature lets the bot open a webview (minimal web page) that can capture information from the user which is not easily conveyed through conversation, such as structured data or a location on a map.

5. Context and memory. These are the two most complicated aspects of your bot. Humans keep track of state and context while making conversation. Bots are therefore required to infer contexts, keep the state of a conversation, and remember key details of previous conversations. This is what differentiates human conversations from most bot conversations these days.

6. Discovery and installation. You need to think about the bot habitat, the listing of the bot in a directory, and ways to initiate the first bot interaction with links and bot affiliation.

7. Engagement methods. These include:

a. *Notifications*—Sending the user new content is a good way to reengage, assuming this is warranted, valuable, and not spammy. In studies done by Facebook the major drivers for engagement with bots on their platform stemmed from bot notifications.

b. *User-led bot invocation*—Providing the users with a way to wake up the bot and initiate a conversation or a task is important, and also something a lot of bot builders forget to add or teach the user about.

c. *Subscription*—Subscriptions or periodic notifications are a great way to keep your bot front and center in the user's life. Letting the users define their interests adds a layer of value to the bot–human interaction (for example, letting the user select interesting topics, to filter the daily news update from a news bot).

8. Monetization. There are various ways that a bot can make money, either directly or indirectly.

This is not an exhaustive list—these are the most common elements, but some bots might have different elements in their composition. Some bots will require interfacing with IoT devices, while others will need text-to-speech and speech-to-text technologies. Some bots can handle tasks across systems and communicate through different channels, requiring a slightly different interface for each chat platform. As a bot designer you should make sure you address the basic attributes outlined here and then think of exploring additional ones.

The design of every user experience stems from the core functionality and purpose of the service or product we are designing. Uber and Lyft are optimized for taking a ride, Google's front page is optimized for search, and so forth. The first step to a successful bot design is understanding what it does. Defining the core purpose and functionality of the bot lies at the heart of your bot's anatomy. Let's do that now.

Core Purpose and Functionality

As a first step in your design exploration, you will need to define your bot's purpose and core functionality. Having a distinct purpose and exposing a particular core functionality is important for every service. This is particularly important with bots, as it is not always obvious or clear to the user what functionality your bot provides. As bot are more limited in the richness of their interface than web or mobile apps, it is important to be very clear about what functionality the bot exposes and provide ways to educate the users on how to invoke that functionality as part of the conversation.

[KEY TAKEAWAY]

As a first step your design exploration, you will need to define your bot's purpose and core functionality.

The bot's conversational user interface means that there are limited ways to remind the users of the things the bot can do. There are ways to solve this—for example, the Google Assistant bot (Figure 5-2) tries to address this issue by offering a "What can you do?" button consistently throughout the conversation.

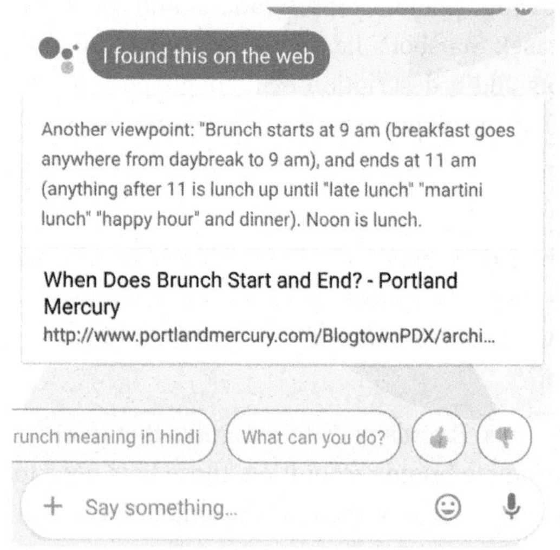

I found this on the web
Another viewpoint: "Brunch starts at 9 am (breakfast goes anywhere from daybreak to 9 am), and ends at 11 am (anything after 11 is lunch up until "late lunch" "martini lunch" "happy hour" and dinner). Noon is lunch.
When Does Brunch Start and End? - Portland Mercury
http://www.portlandmercury.com/BlogtownPDX/archi...
runch meaning in hindi
What can you do?
Say something...

FIGURE 5-2. The "What can you do?" button provides the user with a way to go back to the top menu

Clicking on that button always returns the conversation to the core functionality of the bot, as shown in Figure 5-3.

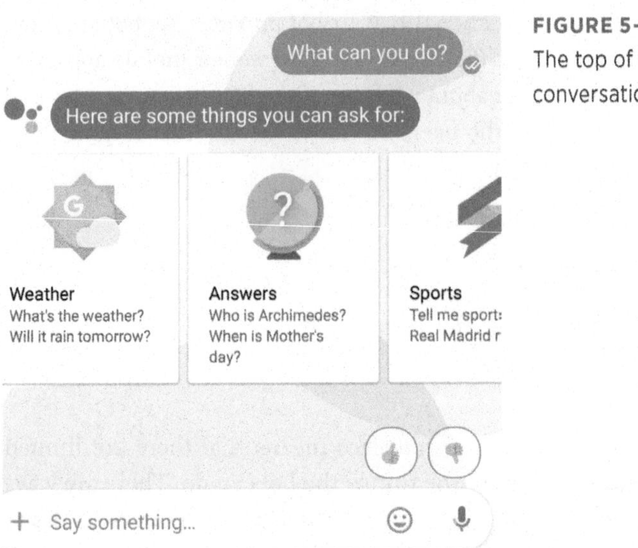

FIGURE 5-3.
The top of the conversation

Other bots surface their purpose in their name. Growbot's name implies growing or boosting (morale, appreciation, and personal growth at work, in this case); Statsbot's name implies providing analytical services. Many bots add a description of their purpose to the onboarding script, as well as the response to the help command.

Closing Thoughts

Currently, we see a lot of bots that serve no real purpose and provide no clear value or set of tasks they can achieve. Many times, when testing bots at Slack, we go back to the developers with the very basic question, "What is your bot good for?"

This is also true for brand bots. Brands need to recognize that in order to provide bots that will promote brand recognition, those bots need to bring value to their users, like we saw in the H&M example in Chapter 2 (Figure 2-1).

Remember, bots are as good as the services they expose. Unclear purpose and lack of usefulness are the number one reasons for bot abandonment and lack of use.

Now that you are armed with a clear understanding of your bot's purpose, it is time to go deep into the different aspects that compose a bot. In the following chapters we will take a deep dive into each attribute listed here, exploring examples and best practices.

[6]

Branding, Personality, and Human Involvement

A man's character may be learned from the adjectives which he habitually uses in conversation.

—**MARK TWAIN**

DESIGNING A BOT INVOLVES a lot of thinking about meta aspects of the interface. In the same way defining a color schema and animations (graphical transformations of elements) takes time in mobile app design, defining the branding guidelines, understanding the personality we want to expose, sorting out the naming conventions we want to use, and outlining the human processes that will support our bots takes time during the bot design process.

Many aspects of a bot's design are tied to the way we want users to perceive our service or product—in essence, we need to figure out the branding of our bot. Let's explore that now.

Branding

Brand management is all about managing how your clients and users in the market perceive your product or service. Branding makes your users remember and love your product, come back and use your service. Branding makes them recognize your bot amongst a sea of others, and even helps with conversions to paying users (users tend to buy more from well-known brands).

Branding includes the names and language you use, the logos and colors, how you provide the service, and in many cases how your service handles situations where things go wrong. Let's discuss a few aspects of the bot that impact its branding and the service it exposes.

VISUAL BRANDING

Whenever you go into an Apple Store, you appreciate that branding is impacted by the way the product is represented visually. Clean and shiny MacBooks presented on clean and shiny tables by a clean and shiny representative. To me it sends a very clear message of "You have entered the shrine of the Big Apple!" The branding emphasizes *quality* and *high-end products.*

There is a misconception that bots do not have visual branding. Yet the conversational UX, a "transparent" user experience, still provides a good amount of visual aspects that impact the branding of your bot—icons, images, colors, and more. Let's use a commerce bot called Kip as an example, and explore different aspects of its visual branding.

[KEY TAKEAWAY]

There is a misconception that bots do not have visual branding. The conversational UX, as a "transparent" user experience, still provides a good amount of visual aspects that impact the branding of your bot.

Logo

Kip has a very distinct logo of a colorful, friendly, and inquisitive penguin. It feels a little like a mascot to me (Figure 6-1).

FIGURE 6-1.
The Kip logo

The impression it creates is one of a friendly and inviting brand. Here are some additional examples (Figure 6-2) of the mascot's usage on the Kip website—the message is one of inquisitiveness and collaboration.

FIGURE 6-2.
The logo is extended into other images on the Kip website

When it comes to the conversational user experience, Kip's logo appears as the profile logo of the Kip bot (Figure 6-3).

Kip BOT 4:42 PM
hello, what can I do for you? Tell me the thing you're looking for, or use help for more options

FIGURE 6-3.
The Kip logo in the conversational interface

The branding is consistent across the bot and the service it provides. Users feel that they are talking to this friendly penguin and, following positive interactions, will associate a strong and positive brand recognition with it.

Using an animal logo was a very smart design choice by the Kip team. This design choice works around issues like gender, race, and other complex branding associations. Remember that users feel like they are "talking" with your company and brand: the bot is a representative of your brand, and is a service extension provided by your brand. The topic of navigating ethical issues of racial and gender bias for service providers merits a book on its own.

Kip was a bot-first company, so they could come up with a single logo that worked for a company and a conversational bot. Some companies, however, already have a set of logos they own and want to use. In many cases, if you have a well-known brand, it is OK to use your company logo as the profile photo of your bot. But remember that you might have multiple bots for your brand in the future, so coming up with a logo that is somewhat differentiated from your core logo might be wise.

Stickers

Kip uses colorful stickers to indicate intent, state, and context (Figure 6-4).

FIGURE 6-4.
Stickers in Kip are functional and reinforce the branding

In addition to their purely functional use, Kip maintains a consistent visual design and branding with its stickers:

- Keeping the penguin logo front and center
- Denoting information with words but also with a visual cue—for example, the "Team Café Cart" sticker has a team of penguins attached to it, indicating the team context
- Using friendly and soft color scheme that extends the friendly brand

Figure 6-5 shows an example of a sticker used in a conversation.

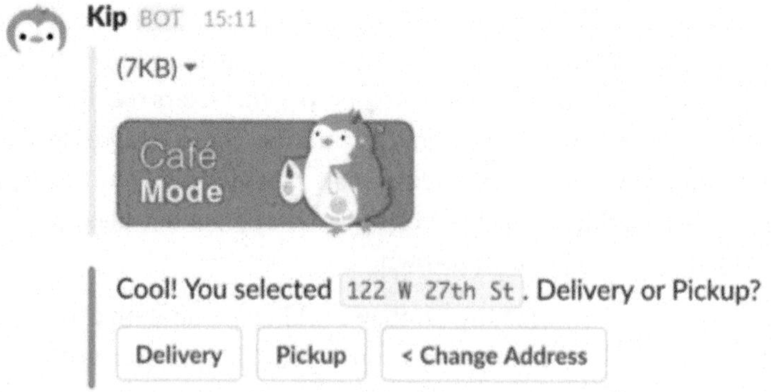

FIGURE 6-5.
A sticker in a Kip conversation

As you can see, the sticker is used as a header to the conversation, setting context but also keeping up the brand recognition throughout the conversation.

Images

Starting from the onboarding, Kip keeps consistent touchpoints using its brand. Images of the penguin are used to explain what the bot is all about (Figure 6-6).

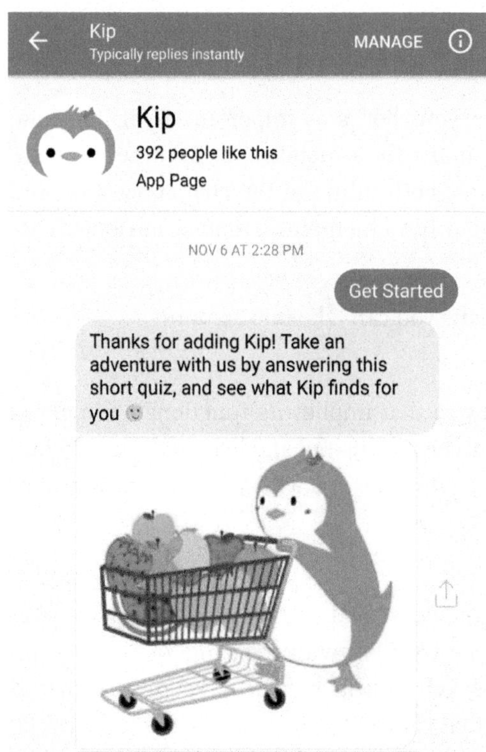

FIGURE 6-6. The penguin appears from the start, in the onboarding

As you can see, the image of the shopping penguin emphasizes the text that describes the functionality of Kip as a shopping bot. Notice that the colors of the fruits are somewhat consistent with the color schema of the stickers we just discussed. Using images with a similar color schema and themes to the stickers makes the design feel consistent and should strengthen the brand association.

The Kip team were very intentional with their visual branding throughout their bot, from onboarding with images, to stickers that reinforce the context and brand recognition, all the way to the logo and color schema. They knowingly over-indexed on visual cues because they were tackling a commerce use case, where brand recognition is paramount.

Not all bots need to be as thorough as Kip when it comes to visual design, but keeping in mind all the touchpoints your brand can have with the user, while keeping a consistent brand throughout, is an important best practice.

NAMING

Picking the right name for your bot is as important as (if not more important than) picking a name for a mobile app. Because there are fewer discovery mechanisms, both inline in the chat application and online on the web, giving your bot a memorable name contributes a lot to usage and discoverability.

There are a few considerations you can take into account:

Functionality
: Naming your bot in a way that implies its functionality (making the name *descriptive*) can be very useful and help make it memorable—Statsbot is a bot that provides you with analytics and statistics in a conversation, for example.

Brand name
: If you have a strong brand name and a single bot that exposes your service, you might name your bot with the same or a very similar name—for example, Lyft calls its bot "Lyft," keeping the brand clear and clean. Note that calling the bot Lyft creates an expectation that it provides the same functionality the ride-sharing mobile app does.

Trademarks
: Try to avoid using another company's trademark. The mobile app market has proven that this is easiest way to get kicked out of a store. Also try to avoid using too generic a name that might conflict with another bot's name.

The name of the bot is not the only naming consideration in designing a bot. For example, bots can add functionality into Slack by means of slash commands. You can think of adding a slash command as like adding an additional command to a command-line interface, or adding options to a menu.

When the user types "/" the slash commands are autocompleted in the input box, as seen in Figure 6-7.

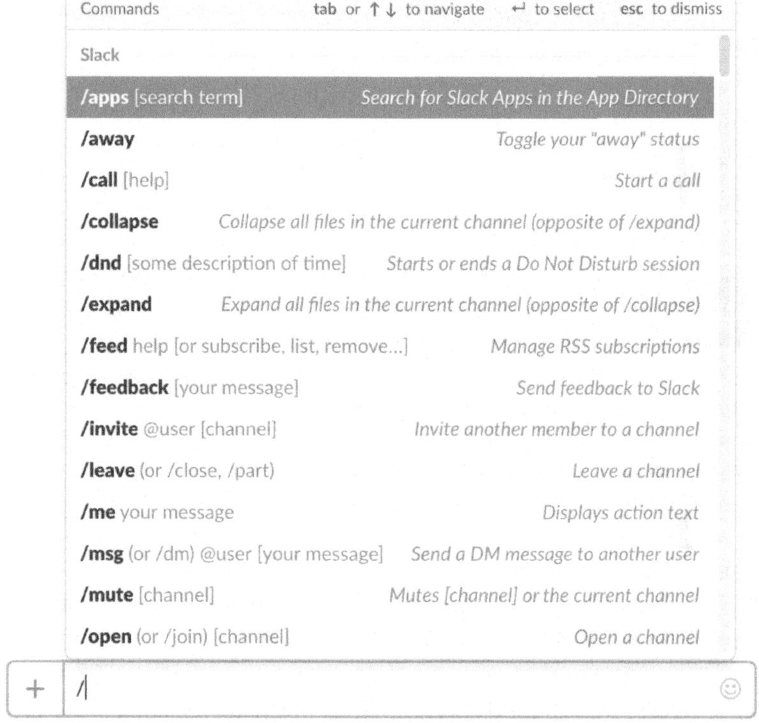

FIGURE 6-7.
An example of a list of slash commands in Slack

Bots can add their own slash commands to this interface. Picking the right name for your slash command contributes to its usage. Slash commands need to be linked to your bot together with the functionality the command provides. It is recommended that you use your brand name as the name of the slash command and use a parameter for the functionality.

Let's take Lyft as an example (Figures 6-8 and 6-9).

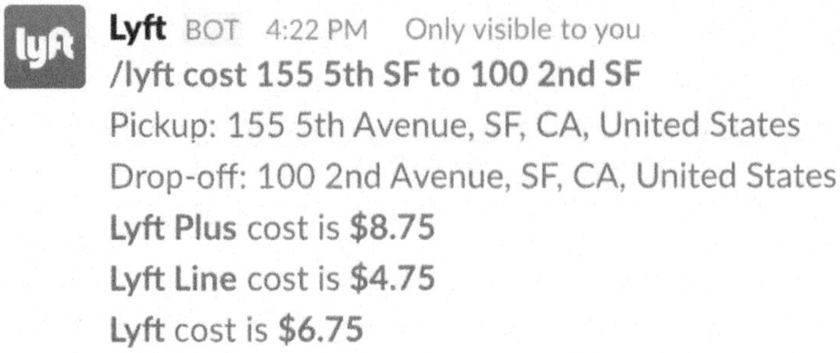

Lyft BOT 4:22 PM Only visible to you
/lyft cost 155 5th SF to 100 2nd SF
Pickup: 155 5th Avenue, SF, CA, United States
Drop-off: 100 2nd Avenue, SF, CA, United States
Lyft Plus cost is **$8.75**
Lyft Line cost is **$4.75**
Lyft cost is **$6.75**

FIGURE 6-8.
A user can type "/lyft cost" to indicate that they want a cost estimate for a ride

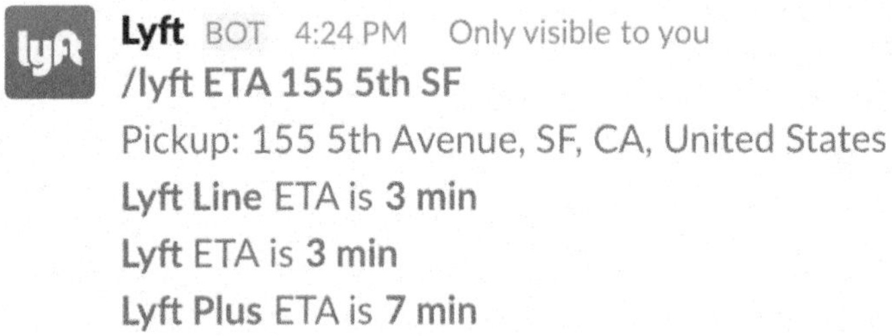

Lyft BOT 4:24 PM Only visible to you
/lyft ETA 155 5th SF
Pickup: 155 5th Avenue, SF, CA, United States
Lyft Line ETA is **3 min**
Lyft ETA is **3 min**
Lyft Plus ETA is **7 min**

FIGURE 6-9.
A user can type "/lyft ETA" (Estimated Time of Arrival) to get the estimated time it will take for a ride to reach the specified address

We will talk more about slash commands in Chapter 9, but for now bear in mind that thinking about the naming of slash commands is another aspect of the bot branding process.

Personality

Personality is one of the key attributes that can differentiate your bot from other bots that provide a similar service. Personality is like the color scheme of an app, or the soundtrack of a movie—something that can provide consistency across the experience and indicate to the users what type of bot they are working with.

We've already seen some examples of how visual branding can help establish personality, in the discussion of Kip in the previous section. In this section we'll take a closer look at the factors involved in deciding on an appropriate personality for your bot, and explore a few examples of different personalities.

There are several things to consider when designing a personality:

Environment
: Consider whether the target environment is a work environment or a consumer environment, and what social attributes are acceptable for a personality in this environment. For example, having a personality that is very humoristic might not be the right choice for a legal assistant bot.

Audience
: Consider the type of audience who will be the primary users of your bot (hint: *everyone* is never the right audience type, even for Google). A bot that talks in slang might not be the right fit for a more conservative audience, and a bot that uses too many three-letter acronyms might miss the mark for others, IMO.

Jobs to be done
: The task the user is intending to execute implies different personality characteristics, even for what initially might seem like similar tasks. Buying a guitar might require a totally different bot personality than buying health care insurance.

Runtime variations
: This is slightly more complex, as it might require some logic associated with the bot, but personality might be context-driven. It is OK to be whimsical when sending directions to a party, but less so when sending directions to a work meeting to which the user is already late.

Locally relevant social acceptance
: Some cultures are different than others. Referring to someone as "dear" might be fine in one place in the world while being culturally unacceptable in another place.

Existing branding

Many brands feel very strongly about the personality their brand exposes. Slack, for example, wants to expose an empathetic, friendly, and pleasant personality.

Values

At the end of the day, the bot's personality is an extension of the service you want to expose. Think about the core values of the service, as that can imply a certain type of personality.

Let's take a look at two examples of bot personalities.

WORDSBOT

The first bot we will explore is the WordsBot, which I created in 2016. Here are the assumptions that led to the personality it currently has:

Name: WordsBot

Environment: Work

Audience: Adults aged ~20–60 using the bot while reading/writing content in English on Slack

Task at hand: Find definitions for words in English

Runtime variations: None

Locally relevant social acceptance: Global safe for work bot

Service branding: The brand is pleasant and productive

Values: Focus on the service, be as transparent as possible

Personality: Simple, getting things done, friendly but succinct, minimalistic, nonintrusive, even dry

Figure 6-10 shows an example of a conversation with WordsBot.

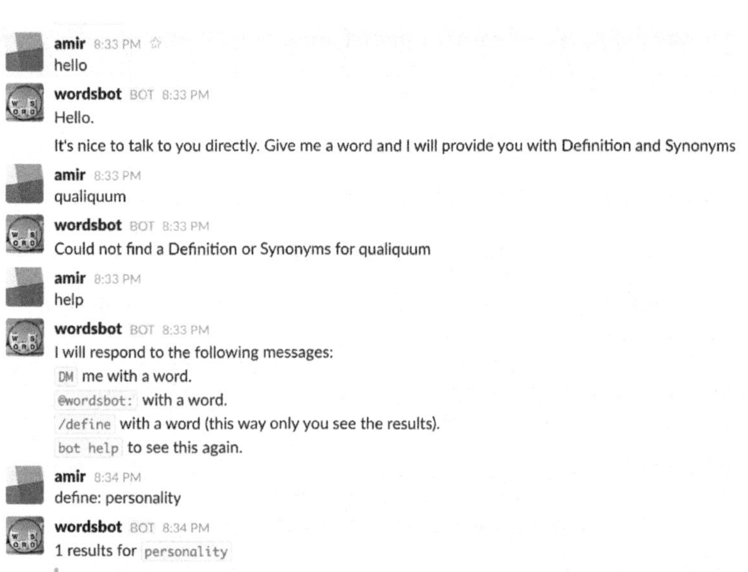

amir 8:33 PM
hello
wordsbot BOT 8:33 PM
Hello.
It's nice to talk to you directly. Give me a word and I will provide you with Definition and Synonyms
amir 8:33 PM
qualiquum
wordsbot BOT 8:33 PM
Could not find a Definition or Synonyms for qualiquum
amir 8:33 PM
help
wordsbot BOT 8:33 PM
I will respond to the following messages:
DM me with a word.
@wordsbot: with a word.
/define with a word (this way only you see the results).
bot help to see this again.
amir 8:34 PM
define: personality
wordsbot BOT 8:34 PM
1 results for personality
Definition (personality)
a person who is widely known and usually much talked about
Synonyms (personality)
big name, cause célèbre (cause celebre), celeb, figure, icon (ikon), light, luminary, megastar, name, notability, notable, notoriety, personage, personality, somebody, standout, star, superstar, VIP

FIGURE 6-10.
WordsBot showing its not-so-shiny personality

As you can see, I adopted a very simple and clean personality here; I did not want the personality to overshadow the service, and I did not see any added value in a big personality. The bot is friendly, functional, and minimalistic, letting the user focus on the task at hand. This is a very common approach for many task-led conversations, where achieving the task is the focus of the conversation, rather than the conversation itself.

PONCHO

Poncho is a sassy weather bot launched as part of the Slack platform and the Facebook Messenger platform. This is my analysis of the bot, based on its personality:

Name: Poncho

Environment: Consumer, fun, social

Audience: Adults aged ~20–40, early adopters

Task at hand: Get weather forecast and notifications

Runtime variations: Errors should be handled with humor

Locally relevant social acceptance: It is OK to ruffle some feathers

Service branding: Fun and humoristic

Values: Get the weather out there and keep it light

Personality: Fun, humoristic, mischievous, comedic, delightful for young people

Figure 6-11 shows an example of a conversation with Poncho.

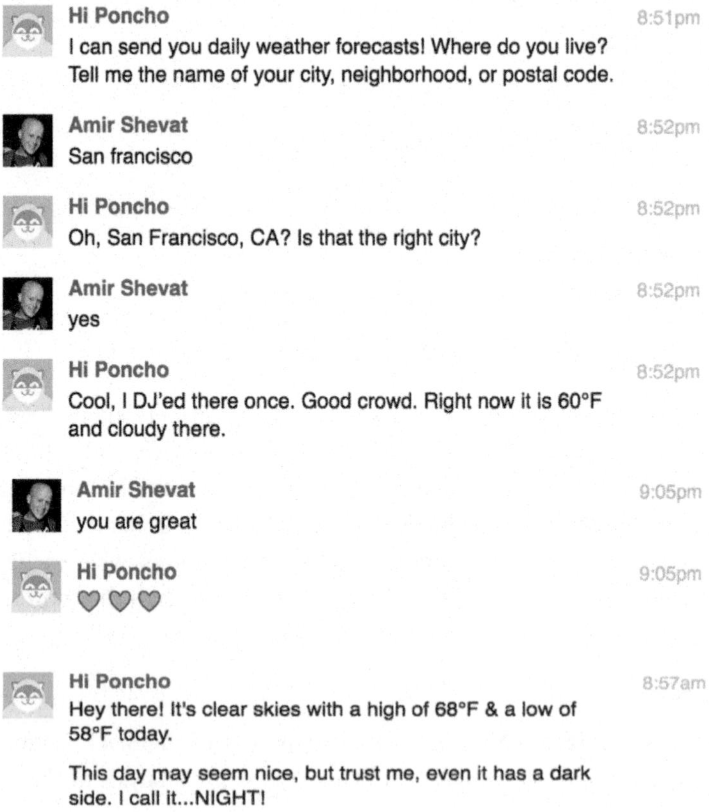

Hi Poncho 8:51pm
I can send you daily weather forecasts! Where do you live?
Tell me the name of your city, neighborhood, or postal code.

Amir Shevat 8:52pm
San francisco

Hi Poncho 8:52pm
Oh, San Francisco, CA? Is that the right city?

Amir Shevat 8:52pm
yes

Hi Poncho 8:52pm
Cool, I DJ'ed there once. Good crowd. Right now it is 60°F
and cloudy there.

Amir Shevat 9:05pm
you are great

Hi Poncho 9:05pm
♡♡♡

Hi Poncho 8:57am
Hey there! It's clear skies with a high of 68°F & a low of
58°F today.

This day may seem nice, but trust me, even it has a dark
side. I call it...NIGHT!

FIGURE 6-11.
Poncho showing its sassy personality on Facebook

As you can see, this is a much more casual bot personality that reflects the whimsical nature the bot builders wanted to relay.

I asked Greg Leuch, head of product at Poncho, about the bot's personality. He mentioned that personality is not just about the text in the conversation:

> We want users to consider Poncho as their friend who tells them the weather every day. Content is written to explain the weather but also to be playful and fun. We also spend time considering pacing. Like comedy, sometimes you need to use timing to land a great joke. Typing indicators can convey to the user that there is more coming. Timing can help ensure the user has adequate time to read a message before sending the next message.

Once you have defined a personality, it is important to keep it consistent across the experience. This gives the users the feeling that they are dealing with a cohesive service (or a persona), which in turn improves trust and engagement.

[KEY TAKEAWAY]

Once you have defined a personality, it is important to keep it consistent across the experience.

Some companies take fun very seriously and hire comedy scriptwriters to spice up their bot scripts. Other bot builders focus on empathy and go into personal questions. Many developers like to keep a clean and minimalistic personality that focuses on task completion while adding little hints to the brand here and there.

Keeping the personality consistent across chat platforms is important in cases where users might use the same bot on different platforms concurrently. Poncho has the same consistent personality in Facebook (Figure 6-11) as it has in Slack (Figure 6-12).

Amir Shevat 2:35 PM
added an integration to this channel: Poncho The Weathercat

Poncho The Weathercat BOT 2:35 PM
So fresh step, so clean

We're looking at overcast skies with temps dropping to 50F later.
Overcast! That reminds me, I have rehearsal tonight with my OutKast cover band, OutCats.

Pawcrafted with ♥ for San Francisco, CA.
(481KB) ▾

FIGURE 6-12.
Poncho showing its sassy personality on Slack

Greg also shared a few insights about consistency in the bot:

> Having a consistently designed voice and personality has been extremely successful for Poncho, and is an important design consideration for any chatbot maker. Take for example a conversation you're having with your friend. If your friend acts peculiar or says things uncharacteristic of them, you know something is off. Like your friend, your bot's voice and character parameters should be understood and clearly defined. These characteristics should not run counter to your existing brand's image. Voice and character become extensions of your brand. They should not respond or handle things that would be counter to the expected characters. And like any company's branding guidelines, the character expectations should be clearly communicated to everyone on your team.

Visual designers spend time on aesthetics, and like them, conversation designers spend a lot of time writing content and functionality that fits the scope and audience of your bot. Knowing how to deal with user expectations, messaging with them in a manner they are comfortable with, and being informative when you can't process a request can make a big difference for the end user's experience.

EXPRESSING YOUR PERSONALITY

Be careful not to overshadow the service you are providing. Just like color scheme, personality needs to be in the service of achieving a task in a delightful way, rather than shining on its own. Some bot builders focus solely on the personality, rather than on the service, and that is like painting a crappy car with shiny colors—it might work for the first impression, but not for much longer than that. A great personality is one that makes your great service shine and keeps the experience delightful and memorable.

Personality can be surfaced in a few ways. Most commonly, personality is exposed in the script itself, for example by adding humor or just picking a set of words that imply a certain personality.

Think of the words that can express affirmation. You can imply a lot about the personality of your bot simply through the words you choose. For example:

Youth, casual: Rad, Amazing Dude!, Way to go!, Boom!

Adult, business: Correct, Affirmative, Great.

You can easily distinguish between the different personalities here based on these simple sets of words. Applying a personality can also mean the use of certain emojis, GIFs, and memes, all based on the culture of the audience and the service.

Deciding to give your bot a personality is not a trivial task, as Dennis Mortensen, the CEO and founder of x.ai, told me:

> x.ai makes an AI personal assistant, a bot called Amy, who schedules meetings for you. Amy is an AI autonomous agent who exists only in dialog with our customers and their guests. She's pure text. We've always believed in the idea of humanizing Amy (and her brother Andrew) and have strategically executed against that from day one. So, we created an entirely new role, AI Interaction Designer, to develop

> Amy's voice and to model the interaction scenarios in order to ensure that Amy delivered the appropriate response in any given scheduling conversation. This role requires a sophisticated sense of psychology and language as well as programming skills since Amy and Andrew's responses are compiled dynamically by the machine based on a plethora of intents and variables.
>
> Once we decided to humanize Amy, it took us down a very specific (and fortuitous) path. For one, we gave our agent a proper full name (Amy Ingram, and Andrew Ingram). And when scripting her end of the dialog, we built in things like empathy. For example, if you have to reschedule a meeting once, that's no big deal. But if you are on the third reschedule, Amy needs to signal that she realizes that this is not an ideal situation, just as a human assistant would. The biggest surprise is how well it worked. People mistake Amy and Andrew for human assistants all the time. They're invited to join calls and meetings and occasionally even asked out on dates.

Dennis also mentioned that building a persona is partly a matter of learning how the human counterpart of the bot—in their case a workplace personal assistant—acts in certain circumstances:

> Our initial reminder logic—those emails Amy would send to guests who hadn't yet responded to her—was honest and fair, but could end up a tad too aggressive in certain scenarios. The goal is not to remind people twice (should they not have gotten back to Amy); the goal is to set up the meeting 100% of the time. Knowing that, we've learned that many of the traditionally accepted social interactions that humans abide by must be catered to in an Intelligent Agent world as well. For example, make sure you don't email again before it is realistic that a guest has seen the first email, however soon the meeting might be scheduled for. Don't send two emails while people are likely to be sleeping, even if the host asked for an early morning meeting. Allow for reasonable lead time. So, don't send me an invite for a meeting at 1:30, at 1:03, which is likely not enough time for people to check their inboxes and get to the meeting.

So what does Dennis recommend to new bot builders?

> If I were to make a recommendation to any aspiring bot entrepreneur, it would be to invest equally in the natural language generation part of the challenge, rather than put all of your resources in the natural language understanding end of the equation.

Dennis has a great point—a lot of developers focus on understanding the user's free text and deemphasize the design of the bot output part of the conversation, thus creating a bot that might be smart, but feels awkward and unpleasant to converse with. Focusing on designing how your bot converses with its users, adding empathy, and making it more friendly and approachable is a great best practice. (We will cover conversation generation in Chapter 16.)

Human Intervention

We cannot talk about a bot's brand and personality without exploring how humans help bots. Bots get their personality from human designers and copywriters, bots can fail over to humans when they cannot handle a conversation independently, and bots might require human supervision to make sure they keep providing the service that the brand offers. In some cases, humans and bot personalities work together to provide a great service. Let's explore that now.

Having humans in the loop is another meta aspect of a bot—bots can help us expose software services, and in many cases can automate mundane and repeatable transactions. While bots can address some use cases very efficiently, having a human in the loop might save the bot from many embarrassing and frustrating situations. In this section we'll take a look at some use cases where having a human in the loop makes sense.

[KEY TAKEAWAY]

While bots can address some use cases very efficiently, having a human in the loop might save the bot from many embarrassing and frustrating situations.

HUMANS RESOLVING AMBIGUITY AND PROVIDING RESPONSE SUPERVISION

In some cases the bot can provide a response to a user, but that response requires human supervision or approval. An example would be a bot that provides the user with a legally binding offer or statement, or when the bot is not sure which of two responses is more appropriate.

Here is one scenario: a human is talking with a sales bot about buying a car, and after a great conversation it is time for the bot to close the lead and offer a financing service to the user. A human sales rep is mandated by company process to manually go over the offer details and approve the loan.

Another scenario is an IT support bot that gets a question with two possible answers. The bot then defers to a human IT professional (providing them with the two potential answers, preferably with a recommendation) and lets the IT professional pick the right answer out of the two.

Note that human supervision has a cost associated with it. Human responses are not instantaneous and will make your bot a little slower. Consider this support email I got from the team at x.ai when I asked why their bot, Amy Ingram, took time to answer my emails:

> We're moving towards a setting where Amy is near instant, but even in her current incarnation she tends to beat most human assistants both in response time (her average time is about 10 minutes) and working days (she works 24/7. No days off. No sleep). All this said, things can get queued up for multiple reasons, mostly due to verification needed or potential response ambiguity (for her).
>
> We operate in a supervised learning environment where we go for accuracy over speed, to ensure high quality in our training data (and product). So a sentence or simple time expression might be pulled aside and that delays things a bit. All while building this verticalized AI.

In Amy's use case human supervision makes a lot of sense. If the response from Amy takes 10 more minutes, but is 10 times more accurate, then that is a price any business user would be willing to pay.

HUMANS ENABLING ERROR/FAILURE ESCALATION

A typical strategy when a conversation with the user is not going well is for the bot to escalate the conversation to a human. There are several common patterns here:

- The bot does not know how to handle the user's intent or request.
- The bot does not understand the user input.
- The bot recognizes negative sentiment (for example, when the user is getting frustrated).
- The bot exposes a way for the user to ask for human assistance and that functionality is invoked.
- The conversation is taking too long, or is unproductive or circular (also known as the user getting lost in the conversation).

While in some cases bots can fail with grace and that is OK, if you are designing for a mission-critical use case, or a use case where failure has a very negative effect that needs to be mitigated, then adding a human in the loop to "make things right" might be a best practice.

HUMANS TRAINING BOTS ONLINE

Another common pattern is where humans help train a bot. As software is great at repeating tasks and at pattern recognition, having humans teach bots by example can be a great way to automate processes on the job.

Let's take an FAQ bot, for instance. The bot gets a question and returns an answer. At first every question is sent to a human for an answer, and the bot acts as a simple router. As time passes the bot learns that certain questions have a distinct, repetitive answer. (What is the meaning of life? The answer is always 42.) The bot then moves to a supervision mode, where it offers the answering human a suggested answer. Once the bot becomes confident in the answer (passing a certain threshold of confidence set by the system), it can answer the question directly without human supervision. As time passes, the bot becomes more and more proficient and requires less and less training and supervision.

This might be a much more effective and agile way of designing an FAQ bot than simply giving it a set of FAQs and letting it fail on every question that is not in the list (especially if you have to anticipate all the permutations of the way a single question can be asked).

[7]

Artificial Intelligence

Intelligence is the ability to adapt to change.
—STEPHEN HAWKING

THERE HAVE BEEN MANY books written on artificial intelligence (AI)—this is not one of them. You can skip this chapter and still design and create bots. Since AI is the technology that underpins bots in several use cases, however, we will provide an overview of common AI services and how they can help you build and design a bot.

Artificial intelligence is commonly mistaken as the essence of bots. Some people confuse and interchange bots with AI. Let's correct this misconception and talk about what AI is and how it can be integrated into a bot to make it awesome.

According to Wikipedia (*https://en.wikipedia.org/wiki/Artificial_intelligence*, as of December 2016):

> Artificial intelligence (AI) is intelligence exhibited by machines. In computer science, an ideal "intelligent" machine is a flexible rational agent that perceives its environment and takes actions that maximize its chance of success at some goal.

Sounds a lot like the way we've described a bot, right? Wouldn't it be great if the service we were building could be automagically converted into this intelligent agent that converses with us like a smart little friend and makes our lives so much better?

The secret to understanding why this is not the case is in the key word of this definition: "ideal."

Currently, we are very far away from this ideal, and while AI has several useful applications, it is far from the intelligent agent of our ideal world.

AI today is not a single thing, but a set of tools that designers and bot developers can use to build conversational bots. AI can help with a lot of the complexities that come with this new interface, but it also adds complexity and a somewhat steep learning curve if you truly want to understand these tools.

[KEY TAKEAWAY]

AI today is not a single thing, but a set of tools that designers and bot developers can choose to use in order to build a conversational bot.

Let's explore a few things AI can do for our bots today.

Natural Language Understanding

Natural language understanding (NLU) lets your bot derive an intent (the need, the intention) from the user's natural language. This process usually involves machine learning with a big dataset of prior conversations (called the *training set*, and sometimes provided with the AI tool) together with specific training and configuration that the bot builder provides. In many cases, NLU is the technology that underpins your bot's conversations.

Most NLU frameworks will help you map (or translate) between user inputs such as "I want to buy a movie ticket," "I wanna get a movie ticket," or "Let's purchase a ticket to the movies" and the intent to buy a ticket, for example.

These frameworks can also extract entities (conversational context variables) out of the user input, such as extracting the name of the theater out of "I wanna go to Cinemax 16," "preferably Cinemax 16," or "Cinemax 16 is my preferred theater." If an entity is missing, the framework can also prompt the user to supply it—for example, it can say "Which theater do you want to go to?" and capture the response to fill in that entity. NLU can also extract context variables and tell you what "it" means following a user input saying, "It has to be *Star Wars*."

Figure 7-1 shows an example of how Facebook's Wit.ai helps you extract entities from the user's input.

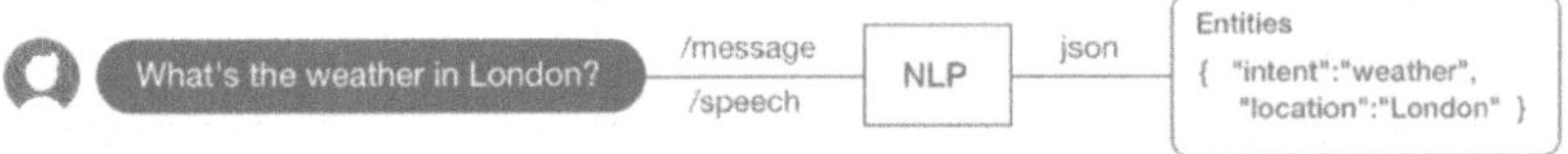

FIGURE 7-1.
Wit.ai converting free text to entities

There are many conversational services that are enabled by artificial intelligence, such as extraction of a date or time, or extraction of a location or address. These are specific use cases that are extremely hard to get right by pure coding.

This complex process does not happen by itself, and there are a lot of tools and services that can help you with natural language understanding. While this book doesn't cover AI in depth, we will show examples of how AI can extract context variables and entities in Chapter 10.

Conversation Management

Extracting entities and intent is an important but basic aspect of conversation management. There is also another layer, which involves managing a multistep flow within a conversation.

Here is how Andy Mauro, the founder of Automat.ai, a company that provides conversational language understanding technology to companies, states it:

> The part of your program that receives extracted intents and entities has to decide what to do with them, and in most cases this means asking the next question, providing the next response or error message, etc. In Alexa or other systems that are mostly "one shot" you don't really worry about multi-turn dialog, yet messaging-based bots are almost ALWAYS based on multi-turn dialog, hence this discipline becoming more important.

Conversation management is a high level of artificial intelligence, in which the AI understands the context of the conversation and knows how to navigate between contexts and subconversations (these flows are sometimes referred to as *stories* in AI solutions like Wit.ai). Humans can easily handle conversational switching, but for machines that is still a hard thing to do.

The SmarterChild bot, one of the oldest bots from the time of AOL Instant Messenger, had a unique ability to pull a user back to a previous part of the conversation with sentences like "Let's get back to X" or "Remember we talked about Y? Let's get back to that." While SmarterChild did not use AI for this trick, conversation management tools might be able to make this transition much easier by finding topics that are related with high confidence.

Intent extraction is hard in a complex conversation too. Humans switch from one topic in a conversation to another with ease. They can talk about a trip, take a deep dive to discuss a restaurant, and go back to talking about the airline service on the way to their destination by just saying "Anyway, the flight there was so wonderful!" Without artificial intelligence helping you extract the right context, it is extremely hard to manage complex conversations in a bot. In reality, human conversation rarely has a single intent—one of the things humans are good at is having conversations where the goal shifts and is discovered over time. The fact that we talk in terms of intents and entities actually shows how far we have to go.

Conversations are also hard to manage because we may need to infer different states of conversational entities (context variables) depending on the context. The entity "credit card" can mean my personal credit card in one conversation topic, but my corporate credit card in another conversation topic. Conversation management should not only help with navigation between one conversation and the other (whether it is a bot-initiated switch or the user initiated the topic switch), but also keeping the right context for the entities for a specific conversation.

Conversation management technologies are in their infancy, and there is a lot of progress to be made in this area. We will talk about the stories and other tools that can help us navigate a conversation in Chapter 16.

Image Recognition/Computer Vision

Image recognition has made a leap of progress in the past few years due to breakthroughs in machine learning. Bots can now, using an image processing service, recognize images, spot emotions in photos, and extract text from images.

This is especially important in use cases where the user uploads an image to the conversation interface and the bot is required to act on the image. Most image recognition services already come trained with a lot of common objects, so no prior training is required by the bot builder. You upload the image to the service and get back a set of objects it recognized, together with a prediction of the accuracy.

A good example of this is the Google Assistant bot in the Allo messaging app (Figure 7-2).

FIGURE 7-2. I uploaded a photo of the amazing cupcakes we had at work, and Google Assistant recognized the objects in the image and offered me conversational topics based on cupcakes

Google actually provides these artificial intelligence capabilities to developers through an API called the Vision API. Figure 7-3 shows another great example.

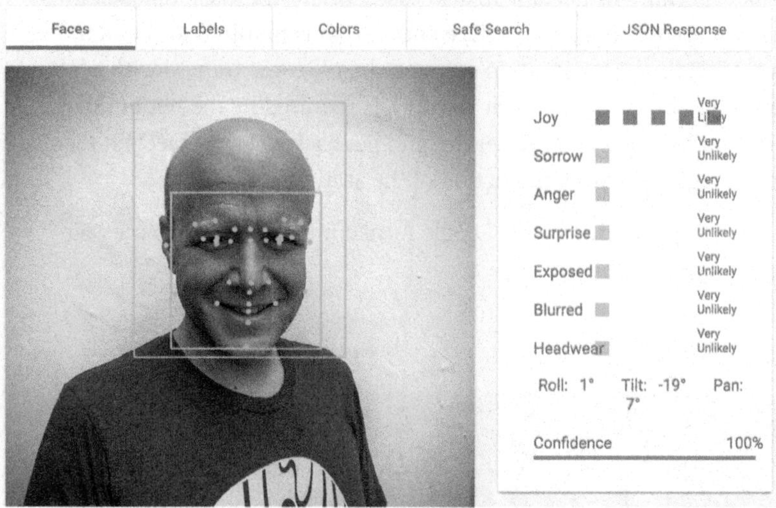

FIGURE 7-3.
Google provides developers with an image recognition API that can be very useful for your bot

Think of a bot that might say something like "What's so funny?"—wouldn't that blow the user's mind?

Machine learning can also be useful for optical character recognition (OCR). This might be used, for example, in cases where a user uploads a receipt or a bill to a finance bot or an expense bot. Recognizing handwriting is more complex and usually yields poorer results than with printed text, but there is progress being made, and with enough training data the outcome can be impressive.

Note that you do not have to support image recognition, and if the user uploads an image in an unsupported use case, the bot can just default to one of the error flows we will discuss later.

Prediction

AI usually does a good job of finding patterns and predicting outcomes based on past data. An example of such a prediction could be the right answer to a support ticket, or the right product to upsell to a user, or the level of confidence that a user will finish a transaction.

Prediction APIs that analyze common patterns can be very useful to bots. You can train these APIs to pick the appropriate responses to specific types of questions that follow these patterns. For example, you can instruct the API that when users ask a question about a particular subject they should get a specific answer, and then teach the bot to pick that answer for future questions like that one.

Another example of prediction (coupled with search indexing) is the Retrieve and Rank API provided by IBM Watson. This is a document search service that uses regular search indexing, like any search service, together with signals from an AI prediction algorithm.

This type of API is not unique to bots, but using a prediction API (such as Google's Prediction API) can help you optimize the conversations and convert non-paying users to paying users.

Sentiment Analysis

One unique AI service that is useful for conversation is sentiment analysis. This AI service gives you a prediction about the user's emotional state, together with a level of confidence in that analysis. Figure 7-4 shows an example of the IBM Watson AI providing an analysis of a Slack channel.

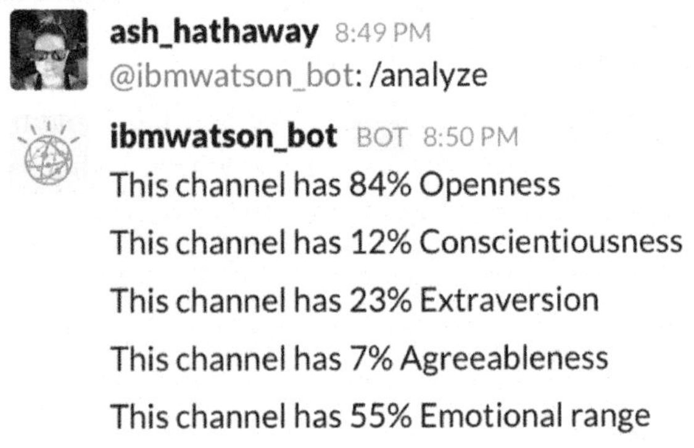

ash_hathaway 8:49 PM
@ibmwatson_bot: /analyze

ibmwatson_bot BOT 8:50 PM
This channel has 84% Openness

This channel has 12% Conscientiousness

This channel has 23% Extraversion

This channel has 7% Agreeableness

This channel has 55% Emotional range

FIGURE 7-4.
The IBM Watson AI picking up different sentiment attributes in a conversation

Why is this important to you as a bot designer? Well, knowing the emotional state of your conversation counterpart has always been key to a successful conversation. As humans we develop empathy, which is a unique ability that lets our minds predict the state of mind of other people while engaging with them. Software has no empathy (at the moment), so using this kind of service can help you build conversational flows that are more likely to succeed, based on the mood of the user.

Please note that sentiment analysis is still in the early stages, and currently not very accurate—catching swear words or terms of endearment through simple pattern recognition might be as effective as AI for many bot use cases.

When to Use Artificial Intelligence

Artificial intelligence provides developers with a great set of tools to use when building their bots. Natural language understanding and conversational management tools make managing complex text-based conversations much easier (compared to the "coding it yourself" alternative). If you are building a conversational bot in a complex domain with the ability to handle pure text conversations, then such AI tools might be critical to the success of your bot.

Using AI for prediction, image recognition, sentiment analysis, and the many other purposes that we have not covered here should be done on a per use case and requirements basis.

Not Using Artificial Intelligence

Not using artificial intelligence is a very valid option when building a bot. Some use cases do not require AI at all. For example, asking a bot to send you news or financial reports every morning might not require AI.

> **[KEY TAKEAWAY]**
>
> Not using artificial intelligence is a very valid option when building a bot. Some use cases do not require AI at all.

AI can be complex, expensive, and hard to implement in a bot. For many use cases AI requires a large training set to be productive. As a bot builder, you can "fake" intelligence for a very long time and still provide a lot of value. At the time of writing of this book, most bots do not use AI. Most bots today use simple regular expressions to understand the intent of the user; they guide the user through rich interactions (giving the user fixed buttons, for example) or use slash commands to mandate structured intent.

That said, AI can be very useful and sometimes critical for building a great bot. Without AI we are limited to specific use cases and somewhat limited value propositions. AI can unlock the promise of a natural conversational interface.

Whether or not you require AI may be a hard question to answer. You can try to fake an intelligent service and see if that works for starters. You can rigidly structure the conversation and sample user satisfaction and error rates. And you can also play with an AI framework and see if you can easily integrate it as a service while building your bot. One input for this decision is the complexity of your script, which we'll touch upon in Chapter 16.

Closing Thoughts

Don't panic! You do not need to learn how to use artificial intelligence right now. Designing a great conversation should be orthogonal to the decision to use AI or any other toolset to build your bot.

[KEY TAKEAWAY]

Don't panic!

In this book we will try to decouple the design choices from the technical choices, while keeping you, the designer, aware of the tools and services that can help you build a great bot.

Next, we will dive into the meat of the principles of designing an effective conversation with users. We will explore the elements of a good conversation and learn from other bot builders what worked for them, and what didn't.

[8]

The Conversation

If it is a ten-minute speech it takes me all of two weeks to prepare it; if it is a half-hour speech it takes me a week; if I can talk as long as I want to it requires no preparation at all. I am ready now.

—PRESIDENT WILSON

According to Wikipedia, human language probably started to develop around 100,000 years ago (in comparison, the first computer was created in 1946). With about 5,000 languages in use in the world today, it seems like humanity should be very proficient at building productive conversations—so why is this still such a difficult problem?

The reason is that until now, humans had to adjust themselves to software, rather than the other way around. As designers we were taught to think in windows, controls, colors, and animation. Without being able to interact with them directly, we were trying to tell the users which buttons to push and which menus hid the information they sought. When designing conversations, we go back to the ancient art that has been at the core of our society for ages: talking to each other.

Remember that bots offer a new interface to an already established human interaction. Bots manifest themselves inside messaging apps, where humans have been communicating with each other for some time. This means that you will need to borrow a lot from preexisting human conversations—for example, users will expect bots not to ignore them, because it is not polite for humans to do so. There is also a big opportunity here to create a connection with your users that is way stronger than what web and mobile apps can ever manage.

[KEY TAKEAWAY]

Remember that bots offer a new interface to an already established human interaction.

In this chapter we will go over different aspects of the bot conversation and explore examples of each. We will practice the best practices established here later, in Chapter 16.

Onboarding

Onboarding is the first interaction users see from the bot—it could be a message that the bot sends to the installing user or a general message to a team. It sets the first impression and tackles a set of tasks that can best be accomplished at the start of the conversation.

In this section we'll take a look at what good onboarding accomplishes.

DECLARING THE PURPOSE

During onboarding the bot declares its purpose in the context of the conversation, making it transparent to the user or the team. The bot should be very clear about what it does and how it can help the user.

Take a look at the introduction from the Howdy bot in Figure 8-1.

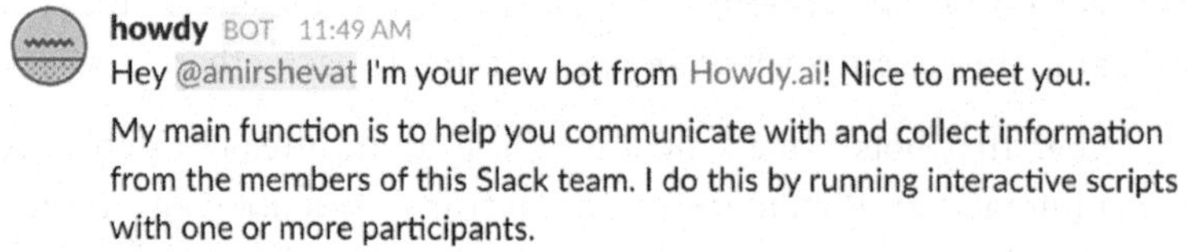

howdy BOT 11:49 AM
Hey @amirshevat I'm your new bot from Howdy.ai! Nice to meet you.
My main function is to help you communicate with and collect information from the members of this Slack team. I do this by running interactive scripts with one or more participants.

FIGURE 8-1.
The Howdy bot introducing itself to the user

You can clearly understand what this bot does—it is a team communication bot. The Howdy development team gave a lot of thought to their onboarding script and how to relay the core purpose and functionality of the bot.

Another example is Poncho (Figure 8-2).

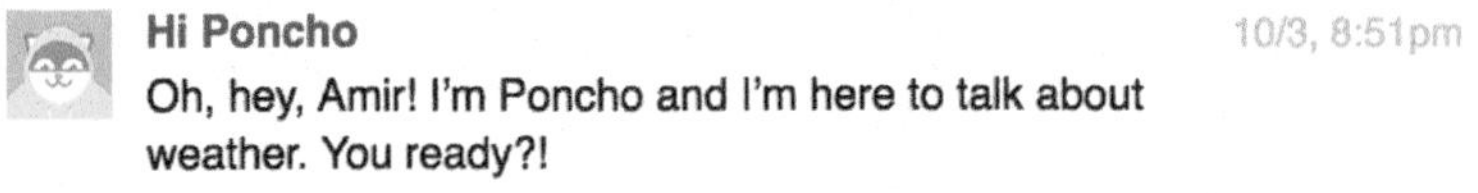

Hi Poncho 10/3, 8:51pm

Oh, hey, Amir! I'm Poncho and I'm here to talk about weather. You ready?!

FIGURE 8-2.
Poncho introducing itself

This bot's purpose is still very clear, although its introduction is not as declarative as the Howdy bot's. Poncho is a much more casual consumer bot, and so a more casual purpose declaration suits it best.

Finally, Figure 8-3 is the first email I ever got from Amy.

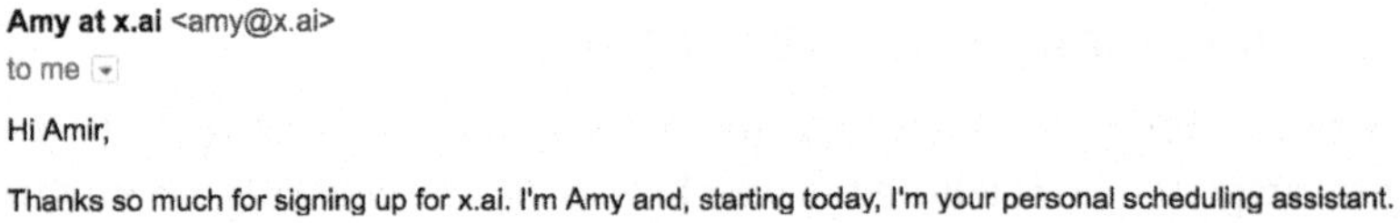

Amy at x.ai <amy@x.ai>
to me

Hi Amir,

Thanks so much for signing up for x.ai. I'm Amy and, starting today, I'm your personal scheduling assistant.

FIGURE 8-3.
My first email from Amy

It could not get more clear than that. As you can see here, the introduction is personal and professional ("starting today" is a work environment term; we touched on this topic when we talked about personality in Chapter 6), and very clear about the purpose of the bot: it is your "personal scheduling assistant." The x.ai team did a great job of clearly defining and scoping the purpose of the bot here—I'm sure it was tempting to just say "I'm your personal assistant," because that wording is more common in real life and might be more appealing to the businessperson receiving this email, but they chose to include the keyword "scheduling" to make the purpose plain.

Stating a clear purpose up front answers the main question most users have today about bots: "What is it good for?" It can be hard to understand the purpose of a bot in a chat interface, as there are none of the visual cues that are common in a mobile app. The minute you get into a photo app, you know what the app is good for because you recognize the images and buttons; you'll usually even find an image of a camera somewhere. A conversation starts with much fewer visual cues.

[KEY TAKEAWAY]

Stating a clear purpose up front answers the main question most users have today about bots: "What is it good for?"

Many bot developers report users trying to ask the bot to perform tasks the bot is not supposed to do. The conversational interface is much less structured than the typical desktop and mobile app paradigm—for example, there is no way to request that the Instagram mobile app tell you the weather (there is no way in the UI to request that service), but with the Instagram-bot this takes just a few clicks ("Hi Instagram-bot, what is the weather like?").

TEACHING THE USER HOW TO USE THE BOT

Now that the user knows what the bot is for, it is time to move to the next stage, which is telling the user how to use the bot. This is a unique point in time where you typically have the user's full attention and can achieve a lot of tasks.

There is a lot of information you might want to relay at this point—the preferred way to communicate with the bot, a way to wake up the bot, the main functions or keywords that the bot supports, slash commands that the bot might expose, and more.

Coming back to Howdy, Figure 8-4 shows the second part of its onboarding script.

howdy BOT 11:17 AM
Hey @amir I'm your new bot from Howdy.ai! Nice to meet you.
My main function is to help you communicate with and collect information from the members of this Slack team. I do this by running interactive scripts with one or more participants.
Since I'm a , I can talk to many people at the same time without getting distracted. I can also wait patiently if someone needs extra time. I will leap to your aid when you command me, and run things according to your channel.
I am so happy that you've given me a shot here on your team. Let's get to work!
What would you like to do first?
These are the key commands you'll need to use my features:
Run
Train
Schedule
Learn More

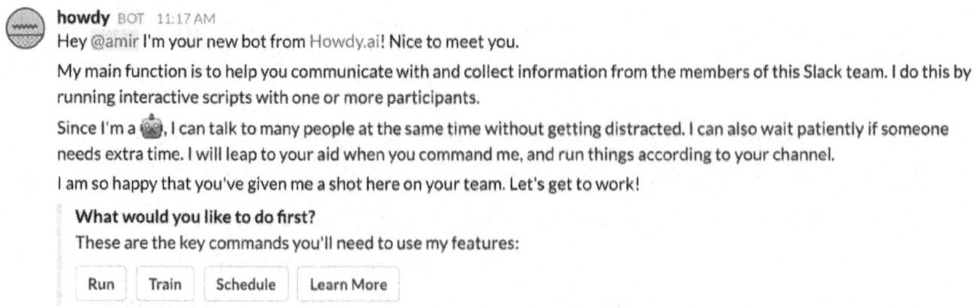

FIGURE 8-4.
Howdy telling the user about its capabilities

As you can see, the bot's designers set clear communication instructions, with the use of actionable buttons (we will cover rich interactions such as buttons in Chapter 9). These buttons make clear what the user can do with the bot next.

Figure 8-5 shows how Amy handles this part of the onboarding process.

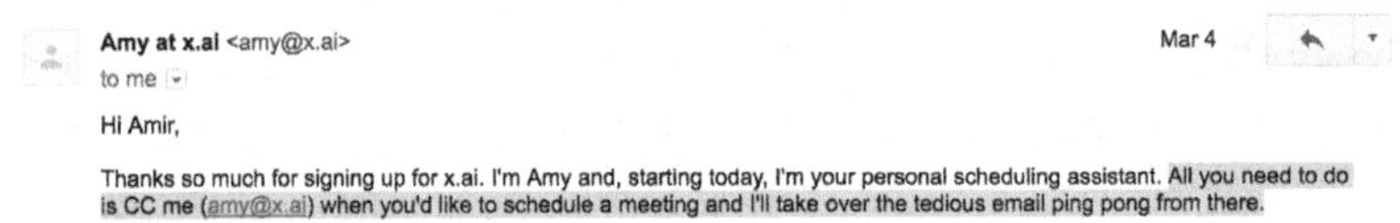

Amy at x.ai <amy@x.ai>
to me
Mar 4
Hi Amir,
Thanks so much for signing up for x.ai. I'm Amy and, starting today, I'm your personal scheduling assistant. All you need to do is CC me (amy@x.ai) when you'd like to schedule a meeting and I'll take over the tedious email ping pong from there.

FIGURE 8-5.
Amy providing onboarding instructions to the user

Amy's onboarding specifies how to work with the bot—CCing Amy on emails will signal to the bot that it is time for it to kick in and take action. As Amy's interaction is passive (Amy will only start working once you email it), it is critical to explain this, or else this will be the last interaction the user will have with the bot. Note that Amy's interaction design is very similar to the way humans interact with each other in this type of situation: when managers need a human assistant to set up a meeting, they CC that person on the correspondence. This makes using Amy very intuitive.

Onboarding scripts can also take the form of a wizard-like interface, similar to in a mobile app, where the bot walks the user through the functionality. Figure 8-6 shows how Poncho does it.

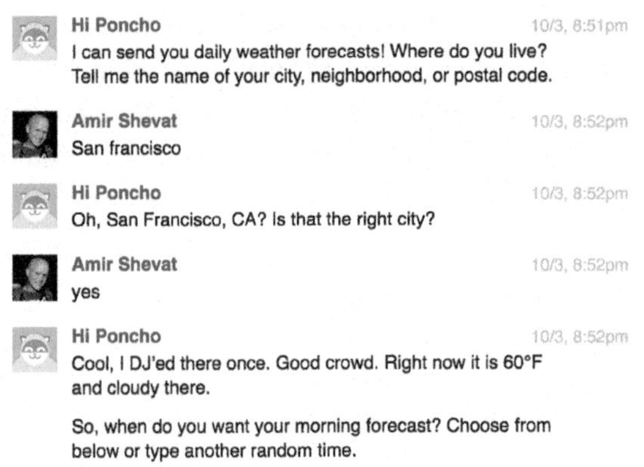

Hi Poncho
10/3, 8:51pm
I can send you daily weather forecasts! Where do you live?
Tell me the name of your city, neighborhood, or postal code.
Amir Shevat
10/3, 8:52pm
San francisco
Hi Poncho
10/3, 8:52pm
Oh, San Francisco, CA? Is that the right city?
Amir Shevat
10/3, 8:52pm
yes
Hi Poncho
10/3, 8:52pm
Cool, I DJ'ed there once. Good crowd. Right now it is 60°F
and cloudy there.
So, when do you want your morning forecast? Choose from
below or type another random time.

FIGURE 8-6.
Poncho uses a wizard-like flow in the onboarding process

Here, the bot demonstrates its value in its initial reply, and then asks more questions. Notice that at the end of this flow the user (me, in this case) is prompted to schedule a daily notification from the bot about weather in the chosen city. This is a great example because it generates hooks for future engagements and also seamlessly helps the user set the configuration of their preferred city.

CONFIGURATION

Poncho's onboarding is a good segue to the configuration part of the onboarding—here you ask the user to supply information that is important to the core functionality of the bot.

In the last section you saw how Poncho asked me to set my preferred city. Figure 8-7 shows how Amy sets the configuration in its onboarding.

To get started:

1. Click here to connect all of your calendars (work, personal, etc.) and set some scheduling preferences. It'll only take a few minutes, promise!
2. Email a friend or colleague and CC me to schedule your first meeting.

FIGURE 8-7.
Amy asking the user to set the calendar configuration as part of the onboarding script

Amy is asking the user to do something a lot of bots need: grant additional permissions to third-party services. In order for Amy to effectively schedule the user's meetings, the bot needs access to the user's calendar. Amy uses the onboarding script to make sure the user provides the appropriate permissions to access the relevant calendars.

This is a common pattern we call *account binding* (or account linking)—the bot knows who the user is on the chat platform, but needs to connect to additional services using the user's credentials. An example of this would be a CRM bot that works with a user in Slack, as in Figure 8-8: the bot needs the user to connect to the CRM in order to act on the user's behalf.

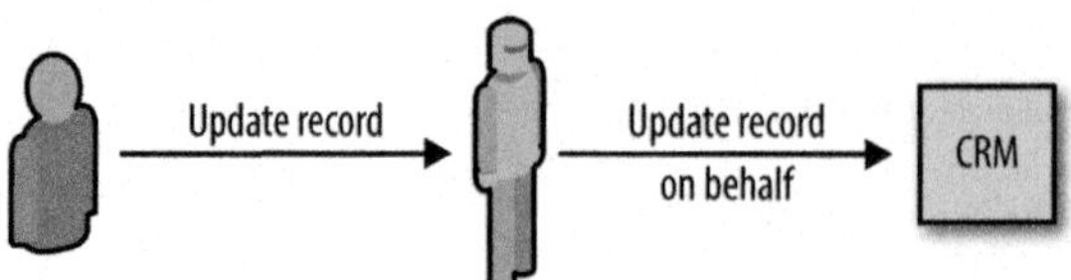

FIGURE 8-8.
Delegated record update—the bot acts on behalf of the user

If the bot does not know who the user is on the CRM system, the operation will fail. This is very common in cases where the bot is connecting with a third-party system—Lyft has the same binding process the first time you want to order a ride through any chat interface.

INCITING USERS TO GET VALUE FROM THE BOT

As you saw with Poncho and Amy's onboarding scripts, it is a best practice to incite the user to actually use the bot as part of the onboarding process.

Figure 8-9 shows how Kip, a popular shopping bot, does that at the end of its onboarding script.

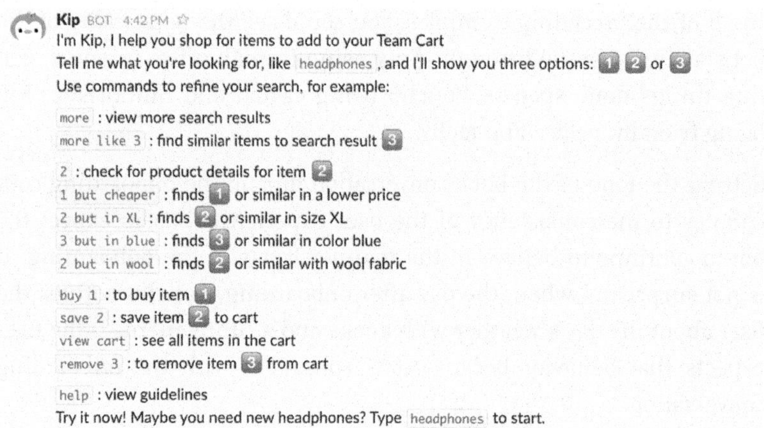

Kip BOT 4:42 PM
I'm Kip, I help you shop for items to add to your Team Cart
Tell me what you're looking for, like headphones, and I'll show you three options: 1 2 or 3
Use commands to refine your search, for example:
more : view more search results
more like 3 : find similar items to search result 3
2 : check for product details for item 2
1 but cheaper : finds 1 or similar in a lower price
2 but in XL : finds 2 or similar in size XL
3 but in blue : finds 3 or similar in color blue
2 but in wool : finds 2 or similar with wool fabric
buy 1 : to buy item 1
save 2 : save item 2 to cart
view cart : see all items in the cart
remove 3 : to remove item 3 from cart
help : view guidelines
Try it now! Maybe you need new headphones? Type headphones to start.

FIGURE 8-9.
Kip's onboarding encourages the user to get started right away

The calls to action we see in Kip, Amy, and Poncho improve the chances of the user experiencing success early in their interaction with the bot. In the same way that shortening the time to "Hello World" for developers has been proven to help get them excited about a technology, shortening the path to "this bot is useful" for users makes them more excited about and inclined to use your bot.

Offering added value to the user at the first engagement contributes to the user's perception of the bot. Useful bots are more likely to be remembered and reengaged with by the user. When I say "useful" I do not necessarily mean utilitarian value; it could be an insight the bot adds or a delightful GIF or witty comment that makes the user laugh. The key is to demonstrate value or provide a meaningful interaction as soon as possible, if not during the onboarding itself.

[KEY TAKEAWAY]

Offering added value to the user at the first engagement contributes to the user's perception of the bot. Useful bots are more likely to be remembered and reengaged with by the user.

SETTING THE TONE AND PERSONALITY

In all of the preceding examples, you could see the personality of the bot starting to shine through. Amy using formal, office-oriented, getting things done speech. Poncho being casual and humoristic. Kip being friendly, geeky, and techy.

Setting the tone of the bot's conversation during the onboarding contributes to the consistency of the user experience—users expect the bot to continue to behave in the manner it has in the onboarding. It is not surprising when, the day after onboarding, Poncho notifies the user about the day's weather with a joke and a funny meme—the user expects that behavior because it is consistent with the onboarding conversation.

Thinking about your brand is also important for setting the tone. If you have a prominent brand it might be useful to use it as part of the onboarding script, in order to create a strong brand association.

ONBOARDING IN A TEAM ENVIRONMENT

In all the examples we've covered up to now, the bot had a 1:1 onboarding conversation with a single user. Now let's talk about onboarding a bot to a new team.

One of the challenges with onboarding a bot to a team is that the team members might not be aware that the bot was invited to the messaging platform, and might not know why they are being engaged by the bot.

Figure 8-10 shows how Sensay mitigates that experience.

sensay BOT 6:54 PM
Hi there! I'm Sensay, a helpful bot that connects you to real people for advice.
A clever admin on your team gave you access right from your DM. 🎉
Let's get to know each other to start off.

Say hi!

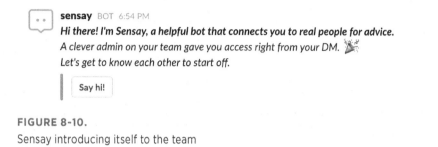

FIGURE 8-10.
Sensay introducing itself to the team

As you can see, Sensay explains that the team admin has installed it and given it permission to direct message (DM) the users. Sensay is a personal bot that, once installed in a specific team, is available to all members of that team.

Being able to DM the relevant users makes sense in this use case (given admin consent). It's important to note, however, that DMing all the users in the team, without explicit permission from an admin, is not advised! I have seen tremendous user backlash because of mass DMing, which is perceived as spamming.

When thinking about onboarding a team bot, think about the paradigm of onboarding a human team member. Let's say a new member joins the team—would you have them DM every person on the team with the same message and start conversing with each person in private? Probably not. Onboarding a team member is a delicate process of finding the right channels to announce the person, notifying the right people who will interface with this person, and making sure that this person is accepted into the team in a friendly way.

[KEY TAKEAWAY]

Onboarding a bot to a team is very similar to onboarding a new human team member.

Bots in teams act as team members, and should be onboarded as such—I usually have my bot ask the admin/installer to add it to the relevant channels, then introduce itself with a short note once it is added to a channel (see Figure 8-11).

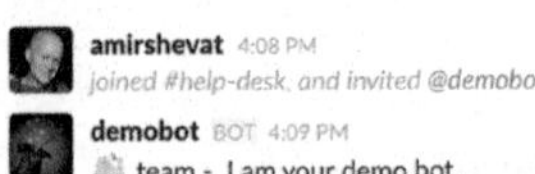

amirshevat 4:08 PM
joined #help-desk, and invited @demobot
demobot BOT 4:09 PM
team - I am your demo bot.
I support direct mentions and DMs, I will read what is in this channel and try to respond accordingly - you can also @demobot: help me.

FIGURE 8-11.
Once added to the channel, it's polite for the bot to introduce itself

Here is a recommended flow for onboarding a bot to a team on Slack:

1. Direct message the installing user (this is an attribute that is exposed to the bot, when it is added to a team) and introduce the bot to them. This should be a regular 1:1 onboarding.
2. Together with the installing user, figure out what is the best way to introduce the bot to the team. Depending on your use case, there are a couple of ways to move forward from here:
    a. The bot could ask the installing user to create a new channel (for example, a help desk bot can create a help desk channel).
    b. The bot could ask the installing user to invite the bot to an already existing channel.
    c. The bot could ask the installing user to send a multi-party direct message to the relevant people.
    d. The bot could ask the installing user to let the bot direct message members of the team.
3. Now that the bot is in the team context, have the bot provide a short introduction describing its purpose and key functionality.

Remembering the exciting paradigm of "introducing a new member to the team" is key to a successful bot onboarding in a team environment. If you have not done so yourself, talk to managers who have done team member onboarding. Also, be domain-specific—for example, if you are introducing a legal bot, talk to an internal legal team manager to see how they introduce team members to the company effectively.

After the bot has been properly introduced to the team, it can communicate directly with team members without the supervision of the installing user—there's no need to further burden anyone. The installing user serves as a hiring manager in this use case, and when the onboarding is completed, the bot is free to act on its own.

Functionality Scripting

As humans, we have been taught for ages to carry on conversations. Most of us have conversations several times a day without thinking about issues like the design or scripting of these conversations. The experience you want the user to have with your bot is very similar: a delightful conversation with a digital friend or assistant. Any exceptions to this guideline will be shortcutting parts of the conversation with rich interactivity, using buttons, images, and other rich elements to relay data visually and capture complex and structured user inputs.

We will explore designing two types of conversations. The first will be a *task-led* conversation, where the target is to accomplish a task. The second will be a *topic-led* conversation, which aims to discuss information and exchange ideas around a specific set of subjects. We have these types of conversations every day—when we buy coffee we have a task-led conversation, but when we discuss a movie we have a topical discussion.

[KEY TAKEAWAY]

There are two types of conversations, task-led and topic-led.

TASK-LED CONVERSATION

The key for designing this type of conversation is finding the optimal set of conversational interactions to complete a specific task. In this section we will give an example of a task and explore a possible flow for accomplishing that task through a diagram.

Let's start by defining a single task—completing this task will be a successful outcome of this conversation. We will imagine building a coffee shopping bot and use the Kip shopping bot for real-life examples. Kip is a complex shopping bot, so we will demonstrate a simple workflow and deal with exceptions in other parts of this book.

Figure 8-12 shows the core task that our shopping bot aims to accomplish. Every interaction will be optimized to get the user to this end.

FIGURE 8-12.

The task to accomplish

Now, let's create a new node in the conversation called *Initiation* (Figure 8-13). There are a few ways to initiate a conversation, and we will cover these in depth later, but for this example we will assume the user initiates the conversation somehow.

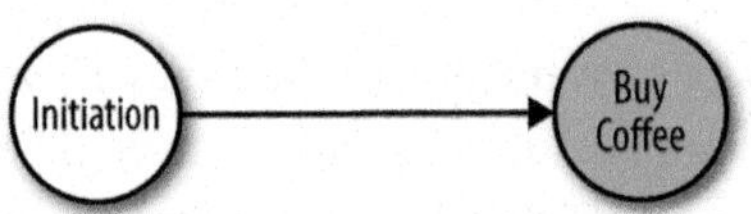

FIGURE 8-13.
Conversation initiation

Now we need to create a flow from initiation to buying coffee. If there is a predefined state (i.e., if the bot "remembers" the user's order), it could be as simple as the diagram in Figure 8-13. For example, I could design a Coffee-bot where every time I initiated a conversation, I would automatically get my cappuccino. I would just need to initiate the conversation with "@Coffee-bot my coffee please!" and the bot would make sure a warm and delicious cup of coffee was delivered to my desk. Wouldn't that be magical?

But life is usually not as simple as that. In most use cases there are several steps the users need to take in order to complete a task. In our coffee bot use case, the user first needs to pick a coffee type. Then, in one flow the user will need to choose whether to add milk, cream, or neither. Lastly, the user needs to provide their location and confirm the transaction. For simplicity, we will assume that the user has a credit line with the bot.

Let's map the initial set of pathways (Figure 8-14).

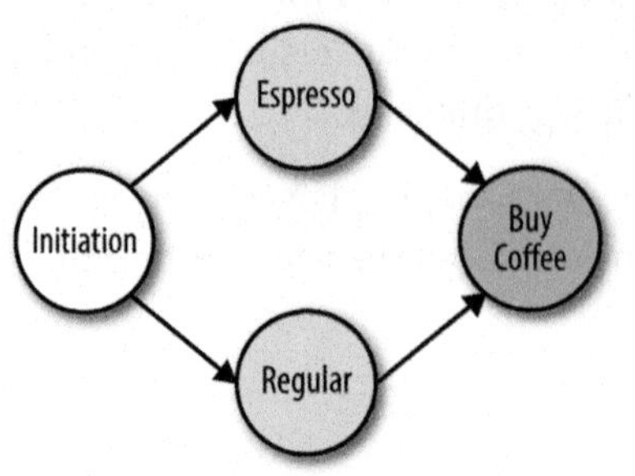

FIGURE 8-14.
The initial paths to task completion

As you can see, one of these pathways might require milk (if you take milk in your espresso, you can stop reading now).

Let's add that option now, in Figure 8-15.

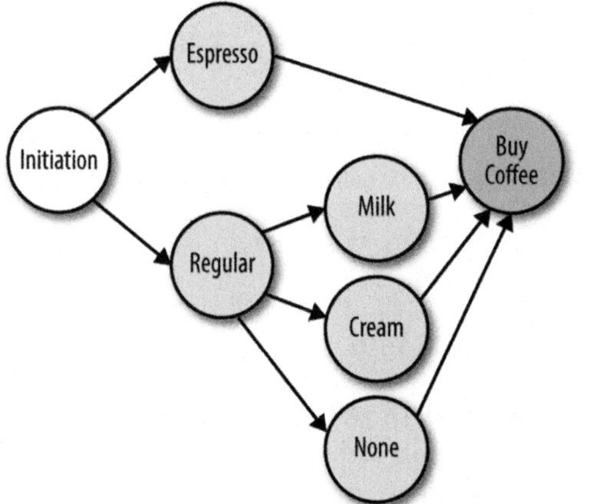

FIGURE 8-15.
Adding an additional step for the second pathway

Now, in Figure 8-16, we will add the address and confirmation stages and complete the full conversational flow.

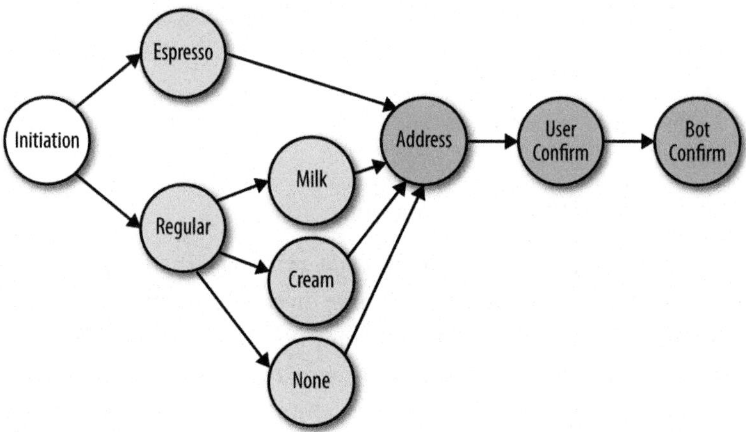

FIGURE 8-16.
Adding the address and confirmation stages

Now let's script the *happy path flow*. This is the most trivial script implementation we can imagine:

User: Hello @coffeebot

Coffee-bot: What would you like today? We have regular coffee and espresso.

User: I will have an espresso

Coffee-bot: Where would you like the coffee to be delivered?

User: 155 5th SF USA

Coffee-bot: Please confirm - 1 espresso coffee to be delivered to 155 5th SF USA.

User: Confirmed

Coffee-bot: Coffee en route, ETA 15 min.

You can see that this is a very optimistic path—the user is providing all the required information on time, in order, and in the way the bot is expecting to get it. If you have ever participated in a user research study, you know that it is really unlikely, in real life, that the conversation will actually unfold in this orderly fashion.

Divergent flows and course correction

Let's explore another flow. We will call this flow *divergent*:

User: Hello @coffeebot

Coffee-bot: What would you like today? We have regular coffee and espresso.

User: Wait, no! I want a cappuccino!

This is a very common situation in conversations—the user has done something you did not expect. They have diverged from the happy flow, and requested something that you were not prepared for. There are two basic ways to handle this: one is *course correction* and the other is *human intervention*. We will talk more about error handling later in this chapter.

Entity extraction

Another flow to explore is called *random order*. Here is an example of how it might go:

User: Hello @coffeebot, can I get an espresso?

In this example the user has provided information in an order the bot is not expecting (this is a very simple case—the type of coffee was entered before the prompt). It would be really irritating if the bot ignored this information, followed the script, and asked "What would you like today?" This is because human conversations are not carried out that way. In a human conversation the order in which you provide the information does not matter.

If we think of key elements of the conversation as entities, and the act of getting these elements of information as *entity extraction*, then we can think of a task-led conversation as a set of data inputs and prompts that facilitate the extraction of these entities.

Here are the entities in our simple conversation:

Coffee type
: A critical and simply structured entity from a small set.

Address
: A critical and complex, less structured entity, from an extremely large set of all the addresses in the world.

Of course, in a real-world scenario, there might be many more entities and types. There could be optional entities such as milk, cream, and sugar; there could be freeform entities such as feedback; there could be critical entities like credit card details, and less critical ones like the answer to the question "How are you doing today?"

In most human conversations, the order in which the user provides these entities is meaningless. I can say I want coffee with milk and two sugars, or I can say I want coffee with two sugars and milk. If you think of the last time you ordered food in a fast food restaurant, you might have noticed that the counter staff were sensitive to the order in which you specified what you wanted. You might have had to order your drink before your food—that is usually due to poor UX design in the point-of-sale systems of these vendors, and it is a source of pain to both vendor and client.

As a conversation designer you need to define and list the set of entities you need to extract from the conversation. You will also need to specify their priority and acceptable data types.

Many artificial intelligence frameworks provide you with sophisticated entity extraction mechanisms: they handle data validation, out-of-order entry, and mandatory entity follow-ups.

If you are used to designing mobile or web interfaces, think about the form paradigm—the user needs to submit a set of inputs, through structured controls like text boxes and select boxes. In a conversation we still need to get and validate all these bits of information, but potentially out of order and sometimes without rich and structured visual controls.

Later in this book we will talk about more structured conversations and richer controls. In many cases you will be able to use these rich controls to extract the required entities without the need to analyze free and unstructured user inputs.

Intent mapping and conversational controls

Let's upgrade our coffee bot to a high-end "San Francisco–grade" barista. Now, users can choose between teas and coffees, between cold and hot; they can pick sizes and flavors. Now the users need to navigate through and filter the choices until they get to the perfect drink.

In a traditional web or mobile app, we would provide a set of controls to navigate and filter the choices. We might have a carousel that flips between drinks. In a conversation, you will need to do that with text.

Figure 8-17 shows how Kip, the popular shopping bot, does it.

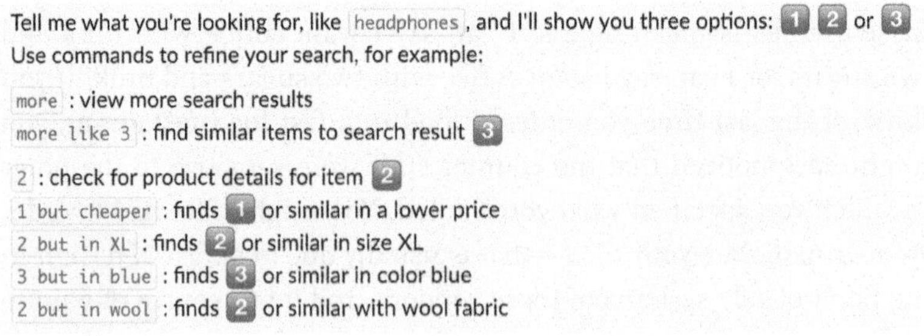

Tell me what you're looking for, like headphones, and I'll show you three options: 1 2 or 3
Use commands to refine your search, for example:
more : view more search results
more like 3 : find similar items to search result 3
2 : check for product details for item 2
1 but cheaper : finds 1 or similar in a lower price
2 but in XL : finds 2 or similar in size XL
3 but in blue : finds 3 or similar in color blue
2 but in wool : finds 2 or similar with wool fabric

FIGURE 8-17.
The Kip bot providing conversational controls to the user

Because Kip deals with extremely large datasets, such as Amazon's inventory, the designers provide a free set of filters and navigation aids to help the user navigate through that potential chaos. As you can see, you can use conversational controls like "option 1 but cheaper" to create filtering.

The key challenge is that users need to remember these conversation controls; they need to remember that they can say "I want this option but in blue." In addition, the conversation itself can become lengthy and cumbersome. Pure text conversation might not be the right design choice for all tasks (we will discuss alternatives in the next chapter).

Another type of navigation challenge is between tasks. Let's say we are designing a travel bot that can book flights and reserve hotel rooms. These are two distinct tasks. In web or mobile apps they might be implemented with tabs or similar top navigation controls. Bots need to provide this *task switching* functionality as well—your bot needs to give the user a way to go back "home," to where they can restart a task or pick another task to execute. This is especially necessary when a user gets lost or stuck while trying to complete a task.

Similar to entity extraction, navigational controls can also be implemented with rich interactions, which will be covered in the next chapter of this book.

Shorthanding

A useful element of a conversational interface is that it can be used to shorthand the complexities of app controls. If the users know what they want, they can just state it and the bot does not need to go through all these entity extraction and navigation steps:

> *User*: @coffee-bot can I please get a short, decaf macchiato with cream and 1 sugar?

Delightful! No need to navigate through endless mobile controls and data entries. Without spoiling the upcoming discussion of context, state, and memory (see Chapter 10), the conversation could even be:

> *User*: @coffee-bot I need my usual coffee!

Now the conversation really becomes useful—the bot becomes a personal assistant that knows and cares for the user, remembers their preferences, and gives them personalized service when they need it. Magic!

Stories/flows

Stories are a way to look at a branch of a conversation. Stories are used to describe a distinct flow or part of a flow. They also allow us to encapsulate or isolate conversational flows. In our coffee bot example, espresso could be one story and regular coffee another. Each story has different elements and possible substories.

The advantage with this approach, from a design and engineering point of view, is that each story can be isolated and possibly more easily communicated between the designer and the development teams. A user–bot conversation can move from one story to another, and if the user diverges from the happy flow, the bot can navigate to a different substory. Substories can include "User wants a coffee that is not available" or "Invalid input from user." There could also be substories such as "cookies upsale" that can be injected into the conversation at the appropriate time, regardless of the current story.

Let's say that our marketing research has shown that people who drink espresso or order milk with their coffee are more inclined to order cookies. Using stories, we can more easily notate this by encapsulating and decoupling the "cookies upsale" story and plugging it in in accordance with our research (Figure 8-18).

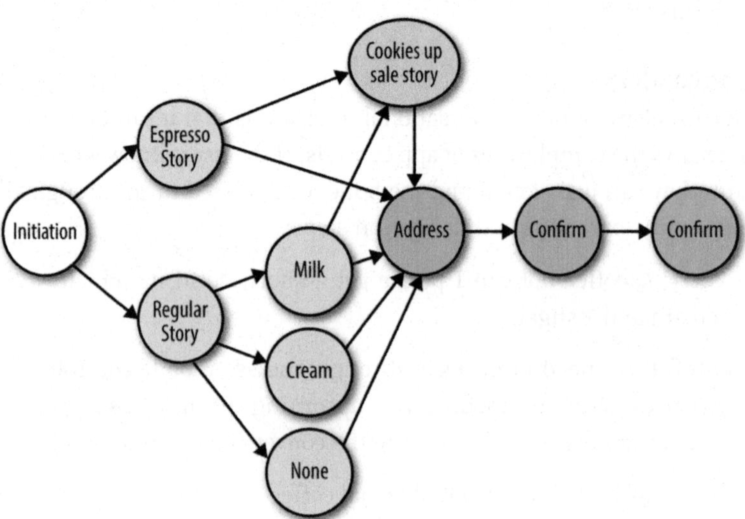

FIGURE 8-18.
Plugging in the cookies upsale substory

This is where artificial intelligence can also come into play. The bot can learn when to effectively use the "cookies upsale" story based on experimentation—it can try to "upsell" and slowly learn the optimized time to do that. For one user being more inclined to order cookies could be correlated with what they've ordered before, while with other users it might be correlated to time of day or even the sentiment of the conversation. The bot can learn and adapt to each user's preferences.

We used a simplified use case of a coffee bot here, but there are many more complex task-led conversation use cases, both in the B2C and B2B domains. For example, think of a legal bot such as the UK-based DoNotPay, which helps you avoid paying parking tickets and also claim compensation for things like delayed plane tickets. The DoNotPay bot could have hundreds of stories that connect together. Decoupling the conversation into distinct scripts makes the bot dialog more manageable from a design and development perspective. We will provide a concrete example of how this is used in a complex scenario, in Chapter 16.

The conversation funnel

Another way to think about and design a task-led conversation is with the notion of a conversation funnel. Funnels are usually used in marketing, when thinking about the user journey through a website or a mobile app. Let's look at this simple sales funnel on a website (Figure 8-19).

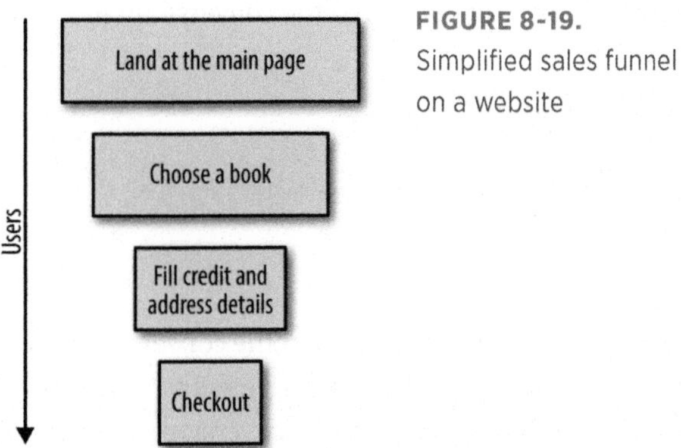

FIGURE 8-19. Simplified sales funnel on a website

We call this a "funnel" because of the basic fact that fewer and fewer users move from one stage to the next throughout the engagement. Users drop off in multiple places in the funnel, and only a fraction of the users end up checking out. A good marketer brings more users to the top of the funnel, and improves the conversion rate at each step of the funnel. A lot of the aspects of a good website design are connected to improvements in the funnel. A larger checkout button can improve the rate of conversion to actual checkouts, for example.

We can create a funnel for our coffee bot as shown in Figure 8-20.

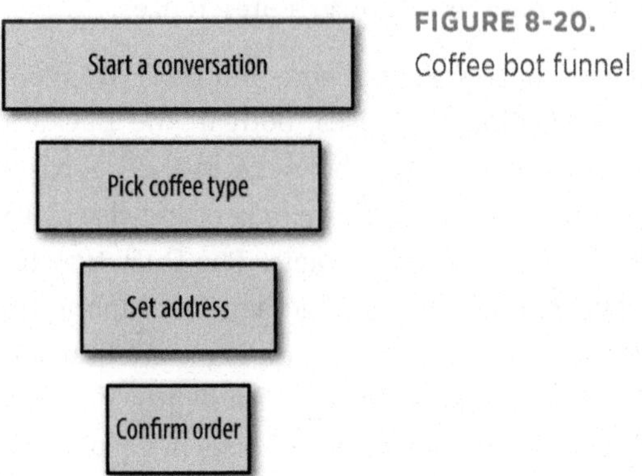

FIGURE 8-20.
Coffee bot funnel

In this example the user goes through a conversation funnel in a very similar way to how they would go through a conversion funnel on a website. Similarly to on a website (where the goal is, say, completing a purchase or signing up for a service), users may drop off and leave the conversation in the middle, quit the conversation right at the end, or just get lost. A good task-led conversation takes the user all the way through the conversation flow and optimizes the conversion rate.

[KEY TAKEAWAY]

A good task-led conversation takes the user all the way through the conversation flow and optimizes the conversion rate.

One thing to note is that the order of the steps in the conversation funnel can be reversed—a user can say "I want coffee with milk please" and only then pick the coffee type. So, an abstract version of any basic conversation funnel might look like Figure 8-21.

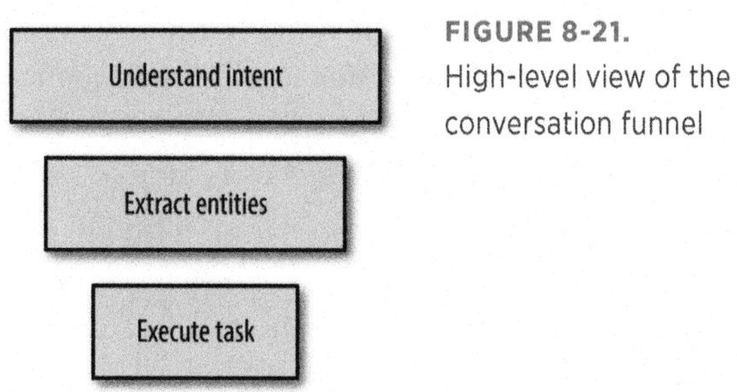

FIGURE 8-21.
High-level view of the conversation funnel

These are the common steps that most task-led conversations follow. Understanding where the users drop off in this conversation funnel is important. By looking at your analytics and logs, you can understand if your bot is having a hard time understanding the intent of the user, is able to extract the necessary entities to finish the transaction, and is able to complete the task. Your job as a bot designer is to optimize the conversation and drive the user through the funnel. This is discussed at length in Chapter 19.

TOPIC-LED DISCUSSION

A topic-led discussion is much less directed than a task-led discussion. It is more circular—the user converses about a set of topics and discusses different aspects of these topics. Think of a task-led conversation as a practical work meeting discussion, and a topic-led conversation as the chat before or after the meeting about what people did over the weekend. A topical discussion does not necessarily have to be casual. Knowledge and learning can also be topical—exploring the world of 3D printing can be a great topical conversation; exploring places to visit in Cancun is another.

> **[KEY TAKEAWAY]**
>
> Task-led conversations need to have the least amount of steps possible to accomplish a task. Topic-led conversations can have more steps, determined by user engagement with the topic.

The key with topical discussions is that you will need to define a set of topics (concrete or abstract) and then facilitate a discussion that circles around these topics. The aim of the bot will be to have a delightful and useful conversation about these subjects.

Concrete topics are topics that are well known at the beginning of the conversation. A good example would be a bot that lets fans talk about a popular movie. For example, Figure 8-22 shows some of the topics fans might want to discuss with regard to the latest *Star Wars* film.

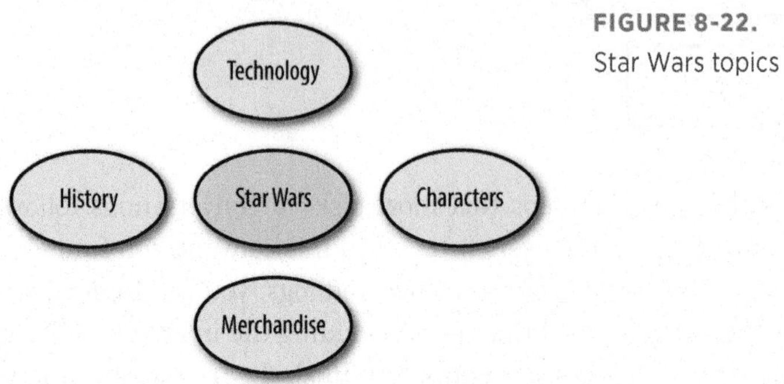

FIGURE 8-22.
Star Wars topics

We can look at each of these topics as a big dataset of information. A discussion circles around these topics and lets the user explore these datasets. You will need to map interesting data objects and attributes for each topic. For example, *Chewbacca* will be an object in the *Characters* dataset, and that object will have lots of different data attributes that might be interesting to surface to the user at different times.

A simple conversation could be as follows:

> *User*: @starwars-bot I am SOOO excited about the upcoming movie!
>
> ***Starwars-bot***: Yes! I am excited too! Did you know that there is a new type of spaceship introduced in this movie?
>
> *User*: No way! I love the star destroyers! Are you going to show those too? And what about Chewbacca? Is he going to show up?

Topic-led conversations are harder to script in some ways than task-led conversations. There is less directionality, and the user can take the conversation down many paths, as well as to dead ends. As you can see in this discussion, the user has already pivoted the conversation from technologies to characters. That is OK! There is a certain amount of user delight in a conversation that is more intuitive and somewhat random when it comes to casual topics. The bot can do this as well, when it understands that the conversation needs to be reinvigorated:

<15 sentences about Chewbacca>

User: Yeah, Chewbacca is great...

Starwars-bot: BTW, did you see the 3-edged sword replica we released to the stores this week? Do you want to see a photo?

User: OMG! Yes!

The bot recognized that the conversation about Chewbacca had come to an end—the user has spent a long time talking about this topic, and all conversational paths around it have been explored. In order to keep the conversation going and keeping it interesting, the bot pivots the conversation to the merchandise domain. We will cover this more in the next section.

From some perspectives, topic-led conversations can actually be easier than task-led conversations, because there is more room for divergence; there is also less of a need for mandatory entity extraction and intent mapping, as the intent and the subject can be abstracted.

An example of an abstract topic could be a search result. The user might be looking something up in the company's knowledge base and discussing that with the bot:

User: @knowledge-bot what is our vacation policy around Thanksgiving?

Knowledge-bot: Company FTEs get 2 days of vacation at Thanksgiving.

User: OK, I also need to know the best person in the company to talk to about PHP

Knowledge-bot: @dana has written 5 internal articles about PHP in our SharePoint site, she might be a great person to talk to.

You can look at abstract topics as a set of concrete topics that are discussed in sequence. But the point here is that, in large datasets, it is hard to build a set of topics and map attributes to these topics. Abstract topics give the user a way to explore and navigate through a large dataset with almost infinite ways to query it. Google is a good example of an abstract topical engine; you can query it on anything and, given enough data, it will know the answer.

In general, topic-led discussions are more oriented toward consumer user cases: window shopping, knowledge and education, media and sports, casual conversation, and more. Business use cases tend to be task-led. Having said that, there are non-task-led conversations relevant to work that can take place around topics such as surveys, user research, peer review, company knowledge exploration, and more.

[KEY TAKEAWAY]

In general, topic-led discussions are more oriented toward consumer user cases. Business use cases tend to be task-led.

Divergence as a way to course correct

In many topic-led conversations, there is no declared task that needs to be completed in a minimal set of steps. The users are not focused on achieving a task; they are focused on exploring a topic. This means that the conversation is less directed and less structured; it is exploratory and sometimes a little random in nature. This in turn means that there is much more leniency toward conversational divergence, and pivoting a conversation is not regarded as an error flow that needs to be corrected. In a topic-led conversation, a bot can even use conversational divergence as a way to recover from input failure:

> *User*: I heard that Chewbacca meets Donatella in the next movie.
>
> ***Starwars-bot***: That is interesting. Did you know you will be able to see Leia again in this movie?

What happened here was that the query of "Chewbacca and Donatella" did not return any results. A bot without pivoting capabilities would have replied, "Sorry, I have no information regarding that," making the conversation more awkward and more likely to end.

Here, the bot pivots the conversation to another character to continue the conversation; it uses a simple "that's interesting" strategy to acknowledge the user and move on. This is a classic strategy we all employ—how many times have you said "that's interesting" and changed the topic of conversation while chatting with someone else?

In the same way, it is expected that the user will change the topic of the discussion on the fly, like moving from Chewbacca to lightsabers, and back to the ethics of the Force. The bot should, in most use cases, entertain these pivots and even encourage them:

> *User*: And these are the 7 reasons I like Chewbacca...
>
> ***Starwars-bot***: Cool. You mentioned you have seen all the movies, which one was your favorite?

The bot's incentive here is to keep the conversation going and the user engaged. Pivoting from one topic to another is a great way to do that. Again, this is a pattern we employ all the time in our personal lives: talking about a topic until we are done with it, and then moving on to another topic.

The trick here is to know when to change topic. You do not want the bot to change the topic too often or when the user is asking a question. Experimentation and machine learning can help you build productive and engaging flows through the relevant datasets.

Entity extraction

Entity extraction can be important in topical discussions too. We do this to explore the user's interests in different aspects of the topic. Figure 8-23 shows how Epytom, the stylist bot, handles it.

So I guess you are ready for the list of 40 pieces?
Choose a section below and we'll pick the best!
Numbers will help you navigate thru all the 40 pieces.
Women
Men
← Back
Write a message...

FIGURE 8-23. Entity extraction in Epytom

The bot needs to extract the user's general preferences; for example, whether they are interested in men's clothing or women's clothing. This is critical information to have in order to have a productive conversation about clothes. Collecting more entities, like favorite color, age, and preferred style, will contribute to more constructive and engaging future conversations.

Intent mapping and conversational controls

We talked about intent mapping and conversational controls when we discussed task-led conversations, but topic-led conversations might also need to rely on intent mapping and conversational controls to answer user's questions and to move from one topic to another.

It is also important to keep entities in context, and let the user change their preferences. For example, when exploring the topic of shopping, I might be interested in men's backpacks for myself, but interested in women's sweaters because I am buying a gift for my wife—the "gender interest" preference will need to change when switching between subtopics.

Stories/flows

The use of stories/flows might even be more important in topic-led conversations than in task-led conversations. Here, you might need to map interesting stories (user journeys) through the dataset. The stories can be connected through association of interest. If a user is using the bot for learning a new language, and the user converses with the bot about what different kitchen utensils are called in Spanish, the conversation might then move from kitchen implements to names of different dishes in Spanish.

Both kitchen utensils and food dishes can be tied to multiple stories, and you can create relations that connect the stories as shown in Figure 8-24.

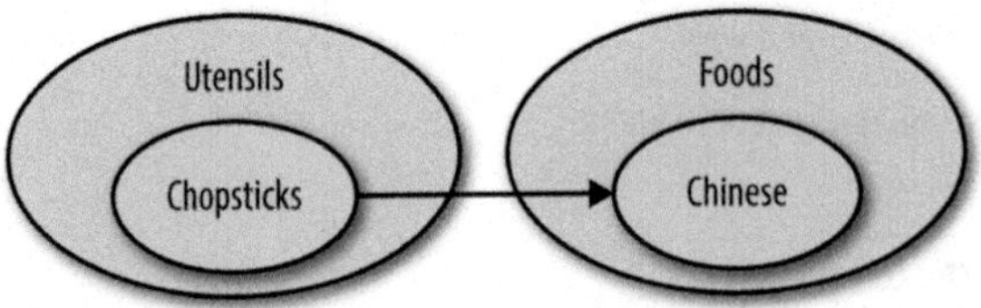

FIGURE 8-24.
Creating links between stories

The conversation can then move from one topic to the other:

> *User*: How do you say chopsticks in Spanish?
>
> ***SpanishLearning-bot***: Chopsticks are called "palillos" in Spanish. Would you like to learn the names of some common Asian dishes in Spanish?

Making these types of associations is done every day in human conversation. The user's mental model connects the two flows, making the conversation more natural and the learning experience more intuitive.

Task-led pathways in topical conversations

In some cases, task-led stories can be intertwined into topic-led conversations. Your bot might be talking to the user about a movie but, when detecting an intent to buy merchandise, transform the conversation into a transactional conversation funnel.

Another example of task-led pathways in topical conversations is prescribed stories within the topic. In our Starwars-bot example, this could be having the bot walk the user through the story line as if they were a hero in the movie. The end task is to complete the journey through the plot, and the bot uses the same tactics of engagement as in a regular task-led conversation.

Now that we have discussed the two types of conversations and explored different aspects of each, let's talk about a few best practices in any bot conversation.

Decoration

Decoration refers to words that we add to sentences that do not contribute directly to the conversation itself, but rather add color and character. There is more to it than just personality—variation adds depth to the conversation and minimizes the sensation that you are talking to a machine.

> **[KEY TAKEAWAY]**
>
> Variation adds depth to the conversation and minimizes the sensation that you are talking to a machine.

Let's take our initial coffee conversation, as follows, and add some decoration to it:

> *User*: Hello @coffeebot
>
> ***Coffee-bot***: What would you like today? We have regular coffee and espresso.

User: I will have espresso

Coffee-bot: Where would you like the coffee to be delivered?

User: 155 5th SF USA

Coffee-bot: Please confirm - 1 espresso coffee to be delivered to 155 5th SF USA.

User: Confirmed

It's pretty dry—it seems like our bot is kinda too serious and also not very friendly. Let's replay this conversation with a few decorations, underscored for your convenience:

User: Hello @coffeebot

Coffee-bot: Wazzap? What would you like today? We have regular coffee and espresso.

User: I will have espresso

Coffee-bot: Awesome sauce! Where would you like the coffee to be delivered? I have the cavalry waiting ;)

User: 155 5th SF USA

Coffee-bot: Got to love San Francisco... Please confirm - 1 espresso coffee to be delivered to 155 5th SF USA.

User: Confirmed

Now it is clear that the bot has a distinct personality and humor; it is friendly and casual. Although the decorations did not contribute a lot of substance to the conversation (other than maybe adding another layer of confirmation), they did add to the flow.

Decorations do not have to be verbal. For example, the sentence:

Coffee-bot: Got to love New York... Please confirm - 1 espresso coffee to be delivered to 155 5th NYC USA.

Could come out as shown in Figure 8-25.

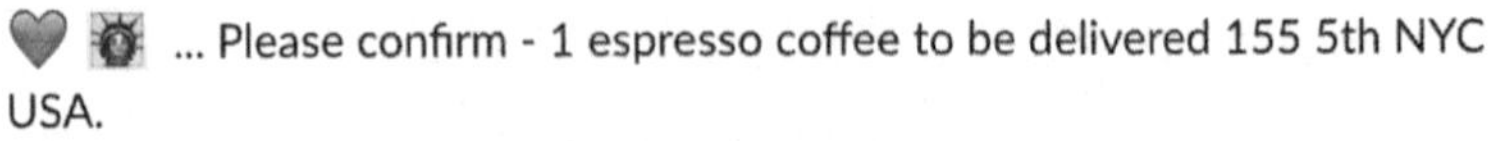

... Please confirm - 1 espresso coffee to be delivered 155 5th NYC USA.

FIGURE 8-25.
Adding visual as well as verbal decoration

With two simple emojis the bot relays an affection for New York City, adding another layer of entertainment to the conversation. We will discuss the usage of emoticons in depth in later chapters.

Another way to decorate a conversation is with memes and images. Figure 8-26 shows how Poncho does that.

Hi Poncho 10/22, 9:14am

Today's forecast: partly cloudy skies with a high of 68°F & a low of 57°F.

For me, it's all about getting *cultured* this weekend. Museums, operas, yogurts...

Hi Poncho 10/22, 9:14am

FIGURE 8-26.
Poncho sends a humorous message and accompanying GIF along with its daily forecast

Poncho adds a GIF that emphasizes the message, encouraging users to look for the funny "extras" that decorate each morning's weather report.

Another interesting use case of decoration is the /giphy bot in Slack—this is a bot whose sole task is to help humans add GIFs to decorate their conversations with each other (Figure 8-27).

/giphy coffee

coffee (2MB) ▾

FIGURE 8-27.
Using GIFs instead of words

/giphy is a super-popular bot, and a source of great satisfaction for users. And this is not surprising—we all love to decorate our conversations. We add facial expressions to scary stories, we reply to threads with funny memes that say more than is polite to say with words. At some companies, like Google, memes are a common and accepted way to express criticism. Emojis and stickers are the compelling value proposition for chat platforms such as Line in Asia, and it seems that this trend is moving west. Whether bots decorate human interactions or their own, adding this layer can make the conversations go more smoothly.

Even if this type of decoration is not built into your bot, it should be able to accept that users may decorate their conversations with it. I have talked to many bot designers who were surprised to see a 👍 from the user instead of a text confirmation. I do not think it is expected that all bots will understand meme sarcasm, but accepting basic emojis might be a wise design choice for certain use cases.

The bot might even prompt the user to use emojis as part of the conversation (Figure 8-28).

Help-desk BOT 7:51 PM
Did that solve your support issue? You can reply with a simple

FIGURE 8-28.
A help-desk bot encouraging users to reply with emojis

Another popular form of decoration on Slack is with text formatting. Adding emphasis (surround your text with *asterisks*) or strikethroughs (surround your text with ~tildes~), indenting the text as a block quote (add an angle bracket to the beginning of your message), and adding code blocks (surround your text with `backticks`) are examples of formatting both bots and users can apply to text.

For example, the following text:

> Hello Team!
>
> *Jon Bruner* is taking a vacation between `12/12` and `12/28`.
>
> > approved by *Tim O'Reilly*

will render a message like Figure 8-29.

FIGURE 8-29.
Text formatting in Slack

RANDOMIZATION

Now let's talk about randomization of outputs. Randomization is another form of decoration, but one that transforms the core part of the conversation. As humans, we do not always use the same phrase to express the same thing. For example, we will not say "I understand" 10 times in a row, even though we want to show that we understand. We will use "got it!" "I see!" "gotcha," and other phrases. Bots that do not randomize their phrases to express confirmation, agreement, or anything else that is repetitive in the conversation tend to sound mechanical and annoying to the user after a while.

> **[KEY TAKEAWAY]**
>
> Bots that do not randomize their phrases to express confirmation, agreement, or anything else that is repetitive in the conversation tend to sound mechanical and annoying to the user after a while.

An example of lack of randomization is the Google Assistant bot, which always says "Yes, I can see images!" when the user uploads a photo to the conversation (Figure 8-30).

FIGURE 8-30.
Google Assistant always posts the same message when the user uploads a photo

While that experience was delightful the first time I saw this message, it got old really fast, and felt a little like a debugging tool that the developer employed to indicate that the bot has received the image. A simple randomization of phrases like "A new image!" "Interesting photo!" and "Another image!" would have made a big difference. People expect randomization. Moreover, it could be a source of delight—not knowing what the bot will say makes it more interesting to users. The fact that every time my kids ask Alexa to "tell me a fact" it spits out a new piece of information makes the interaction very engaging for them.

Keeping the conversation fresh is something that will delight your users and improve the chances of reengagement. Users will be waiting to see what the bot will say next, in the same way they will be inclined to interact again with an interesting person with fresh new ideas and thoughts.

Priming the User to Give the Right Information

One of the challenges in conversational interfaces is handling user inputs. Users can provide the same information in many different forms. For example, denoting April 3, 2017, might be done in several ways:

- April 3rd
- 3rd of April
- 3 April 2017
- April 3 2017
- 03/04/2017
- 4/3/2017
- Next Monday
- First Monday of April
- Today
- Tomorrow

These are all valid ways to say the same thing. The team at x.ai spent a lot of time on understanding the date the user is referring to. Some bot builders use artificial intelligence to solve this problem, with some success.

There is a way to optimize the conversation and drive the user to say the right thing (in the right format), and it is called *priming*. Getting the right answer depends a lot on how you frame the question. Instead of saying, "When would you like the meeting to take place?" the bot can say, "At what *date* would you like this meeting to take place?" This primes the user to provide a date format rather than saying something like "The day after tomorrow."

While this is not a bulletproof solution, it lowers the error rate and minimizes the compensation efforts your service will need to make, and also provides an easy way to streamline the learning process and educate the user on how to use the bot.

[KEY TAKEAWAY]

Getting the right answer depends a lot on how you frame the question.

Another priming strategy is to limit the options the user has, improving your bot's chances of predicting the response and understanding the user.

Here is an example:

> ***Bot***: I am ready to send the meeting invitation for your meeting on April 3rd at 2 p.m. Would you like me to send it now or modify?

The funny thing is that a lot of users would answer "Yes" to this question. This sounds like a yes/no question, so the user picks one. Unfortunately, "Yes" is not the answer we are looking for; moreover, it is an ambiguous answer that the bot does not know how to interpret.

Here is another way to say the same thing:

> ***Bot***: I am ready to send the meeting invitation for your meeting on April 3rd at 2 p.m. You can send or modify. Which one would you like?

Now answering "Yes" makes less sense, as this is does not look like a yes/no question. Some users might still do that, but it's much less likely. There may even be a better way of encouraging an appropriate reply, with visual hints:

> ***Bot***: I am ready to send the meeting invitation for your meeting on April 3rd at 2 p.m. You can **send** or **modify**. Which one would you like?

Making the options bold primes the user to pick one of these options specifically. There are other priming options you may be able to use, and we will discuss these in the next chapter of this book, but even without these, a simple word choice or some basic formatting can prime the user to do the right thing.

It is also important to consider user expectations, as Vittorio Banfi, cofounder of Botsociety.io, shared from his experience (Botsociety is a bot-designing tool):

> Some bots' designs have a problem of aligning the user expectations with the bot's purpose and capabilities. For example, for a train ticket booking bot, it is wrong to ask "Where do you want to go?" This question does not align the user expectations with the bot capabilities. For example, the user is not able to understand immediately if the bot expects a city, an address, or even something more personal, like "I want to go home." Unless your bot is capable of processing all of those phrases correctly (and it probably isn't), then you will need to design its conversation better, by aligning the user expectations with the bot capabilities. For example, a far better question would be:
>
> *@bot*: Where do you want to go? You can say for example "I want to go to Austin."

Limiting the options for the conversation to derail and applying simple priming techniques can significantly improve your bot's usability and the experience users have with it.

Acknowledgment and Confirmation

A specific set of common use cases involves user input or process confirmation. This is where the bot acknowledges the reception of an input and confirms its correctness. It also includes a bot asking the user for confirmation before executing a specific process.

Let's take a look at a few design principles that should guide this aspect of the conversation.

RESPONSIVENESS

A bot should never ignore a user—when a user asks a question or makes a comment, the bot should reply to the user, either with an acknowledgment or with a related sentence. This is true for all input types, even if the bot is not designed to handle them (for example, if the user uploads a photo to a text-only bot). The bot should not ignore the input—a simple "I don't handle files and images" might suffice.

[**KEY TAKEAWAY**]

A bot should never ignore a user.

Chitchat is also an aspect of responsiveness. Here is an interesting fact shared by the team at Dashbot, a bot analytics tool:

> 72% of bots receive "hi" [Figure 8-31]—so, remember to have a good response to this greeting. This is a great opportunity to reinforce your brand and the bot's personality.

case sensitive	
hi	65.7%
Hi	57.6%
hello	54%
Hello	42.3%
👍	40.5%

case insensitive	
hi	72%
hello	60.3%

FIGURE 8-31.
Percentage of users greeting the bot

> About 12 percent of our Facebook bots have had users ask the bot to tell a joke. Easter eggs are a great opportunity to both build a personality for your bot and "surprise and delight" users.

Many bot builders also report that users tend to thank their bot, or tell them they love it. Simply acknowledging these types of input is important for ongoing engagement with the bot.

Another aspect of responsiveness has to do with the amount of time the user needs to wait to get a reply. If a user's request takes a long time to process, the bot should indicate that it is working on the request. This could be implemented with a simple "working on it" type of message, or the "typing" indicator provided by some of the platforms. Just letting the user know that the bot is processing the request will prevent the inherent fear of "the bot is not working" that a lot of users report.

In very long-running processes it might be useful to give the users an indication of when they can expect the results. I have seen a bot that said "I will have the report ready shortly" and actually posted the report 24 hours later. Remember to set expectations right and deliver on them.

EXPLICIT VERSUS IMPLICIT CONFIRMATION

Getting confirmation from the user is an important step in many workflows. There are two types of confirmations in a conversation: *explicit* and *implicit*.

Explicit confirmation typically involves checking with the user that the input provided by the user was processed correctly, or requesting permission to act. With explicit confirmation, the bot will not perform the action until it gets the confirmation.

Here are some examples of explicit confirmation:

> ***Bot***: I will set your preferred address to be 155 5th Street, San Francisco. Did I get that correctly?
>
> ***Bot***: I will move the meeting to April 3rd pending your approval. Please confirm.

Implicit confirmation, on the other hand, confirms that the input has been received or that an operation will take place *without* asking for user approval. Think of it like a personal assistant sending you an FYI notification.

Here are some examples of implicit confirmation:

> ***Bot***: I will set your preferred address to be 155 5th Street, San Francisco.
>
> ***Bot***: I will move the meeting to April 3rd.

Some use cases require explicit confirmation. For example, if the bot is facilitating a bank transfer, it is very wise to confirm the critical details. Other use cases do not require an explicit confirmation, like picking a favorite color in a game run by the bot.

Explicit confirmation is very taxing on humans. We do not need to confirm every aspect of our conversations with our conversational counterparts. Use explicit confirmation only in use cases that mandate it, or if you are not confident that you are processing the user input correctly. I recommend that you default to implicit confirmation when the bot is confident of the user input and in "softer" use cases where the cost of getting it wrong is not very high.

Avoid repetition

If you can't avoid having confirmations throughout your bot conversation, make sure you follow the randomization principle discussed earlier. It is really annoying to have a conversation with a bot that keeps saying: "Do you confirm X?" "Do you confirm Y?" Keeping the conversation natural by randomizing the way the user confirms the input or action is highly recommended.

This is also true with error messages. Consider having multiple fallback messages and rotating them so that any misunderstanding doesn't seem like the equivalent to an HTTP 404 (Not Found) error. Nobody wants to hit a dead end, but if your fallback responses are creative enough, then you might be able to defuse the situation by delighting users with the unexpected.

Accept user confirmation permutations

Not many things are more aggravating than needing to confirm the same thing many times. Make sure you accept confirmation in all its permutations: "Yes," "Confirmed," "OK," "Y," "YES," and all the many other ways to say "I confirm."

If you can accept batch confirmations, all the better. Take for example the Amazon shopping flow—the user gets a page at the end of the flow with all the details bundled together for confirmation. Your bot can do the same:

> ***Bot***: We will deliver X, to your address Y, at Z - confirm or modify?

As you can see, the bot is confirming a bundle of items that compose a complete transaction, collapsing three separate confirmations into one.

Consistency

Conversation design should be consistent and thoughtful throughout, in the happy flows and in error flows. In the same way you would strive to keep a consistent look and feel across the pages of a website and the screens of a mobile app, you need to keep a consistent experience in every aspect of the conversation.

[KEY TAKEAWAY]

Conversation design should be consistent and thoughtful throughout.

Bot designers are sometimes not involved in designing the feedback section or the error flows, leaving these sections unstyled. Users get surprised by—and perceive negatively—bots that provide a delightful experience until something goes wrong.

Here's an example of inconsistency:

> ***@travel-bot***: I would be delighted to book your flight. Where would you like to go?
>
> *@user*: I wanna go to SxSW
>
> ***@travel-bot***: Error finding sxsw, please re-enter input.

In this example, the bot started out as a friendly conversational agent, and turned into a not-so-approachable bot when an error flow occurred.

Consistency should also be maintained when the conversation is routed to a human supervisor. Humans managing the conversation should be aware of the branding and the style the bot provides as the interface of the service, and keep the conversational tone consistent.

Reciprocity

Reciprocity is a key aspect of human interaction. Every conversation is composed of reciprocal give and take. Understanding reciprocity can help a lot when designing a productive conversation. Here are a few strategies.

COMMUNICATE VALUE BEFORE ASKING FOR INPUT

Users are willing to invest a lot in apps—taking photos, tagging friends, adding location details, describing preferences, and even giving credit card information—as long as they understand what they are getting in return. One of the common mistakes in designing a conversation is to forget to communicate the value to the user:

> ***Banking-bot***: Thank you for installing me. Let's add your first bank account - enter your Citibank account number please.

While this bot is friendly and polite, there is no communication of value, and therefore less incentive for trust and willingness to provide this super-sensitive information. Making some slight modifications to the conversation might improve that:

> ***Banking-bot***: Thank you for installing me. I am your Citibank bot: as such, I will notify you of upcoming bills, send payment alerts, and let you do balance checks at any time.
>
> ***Banking-bot***: Let's get started and add your first bank account - enter your Citibank account number please.

Now the user has a clear understanding of the value associated with the bot. There might even be a "Wow! I want this" moment that will help build trust and excitement about engaging with the bot. Users will potentially then be more inclined to provide sensitive information and cooperate with the bot.

INITIATE AND REVIVE ENGAGEMENT WITH QUESTIONS AND OFFERS

Facebook reported, at one of its recent developer events, that a high percentage of user engagement is initiated by bot engagement. Having your bot ask a question or offer an action kick-starts a reciprocal engagement. Humans have been trained to answer questions and accept/reject offers—it is considered impolite to ignore such things.

On the other hand, you need to make sure the questions or offers are relevant and timely. Because this is an intrusive action, the user might get the feeling that the bot's interjections are spammy or irrelevant. A good way to make sure you have consent, and make your intent clear, is to get approval from the user for these interactions:

> ***Deal-bot***: Would you like me to send you the daily deals?
>
> ***Expense-bot***: Should I send you a reminder to do your expense report on the 25th of each month?

Now the user knows to expect scheduled alerts and offers, and is aware of the value associated with these.

PROACTIVITY

Related to the idea of reciprocal conversation is the mutual proactivity of all parties in the conversation. Humans will expect, and in most cases positively respond to, a timely suggestion from the bot.

[KEY TAKEAWAY]

Humans will expect, and in most cases positively respond to, a timely suggestion from the bot.

Figure 8-32 shows how Statsbot, a marketing bot for work, offers the user to do another segmentation analysis.

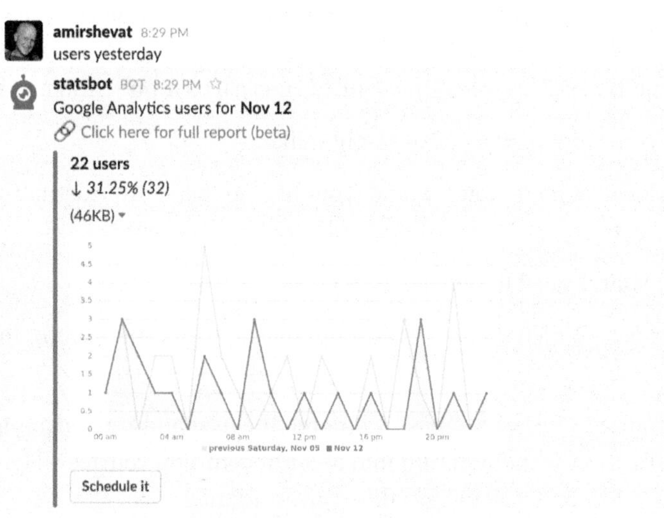

FIGURE 8-32.
Statsbot engaging proactively with the user

The bot actually "upsells" the next interaction, educating the user on the go about other (or new) features of the bot and improving the chances of future engagement with the bot. The key to an effective proactive engagement is doing it in a timely manner, connected to the user's original intent. Providing the user the ability to subscribe to a notification of the same metric (as you can see in the button in Figure 8-32) is another example of proactively engaging the user and promoting further interaction. Timing is tricky, but being proactive after they've experienced success or achieved a target is a good way of engaging the user.

Some designers are afraid of this type of interaction, considering it pushy, but if done right, this can be a great way to improve the value of the overall conversation. You will be surprised what users will do if you just ask them nicely and at the right time.

We will cover this topic in more detail when we discuss bot engagement methods in Chapter 12.

COMMON COURTESY

There are a lot of reciprocal behaviors in a good conversation that we as humans might refer to as common courtesy—thanking someone for doing something, acknowledging your counterpart's inputs, and more.

Here is a common mistake a bot might make:

> ***Travel-bot***: Which hotel would you like to book in Austin for TalkAbot?
>
> *User*: Wait, I need to rent a car first.
>
> ***Travel-bot***: Which hotel would you like to book in Austin for TalkAbot?

The frustration! The bot completely ignored the user and kept on going in the conversion without taking into account common courtesy. Here is a slightly different way to have this conversation:

> ***Travel-bot***: Which hotel would you like to book in Austin for TalkAbot?
>
> *User*: Wait, I need to rent a car first.
>
> ***Travel-bot***: Sorry! I'm not capable of renting cars, but I can help you book a hotel room. Which hotel would you like to book in Austin for TalkAbot?

Now the bot has acknowledged the user's request, expressed empathy, and continued the conversation. While the user did not get exactly what they needed, the conversation is much less frustrating.

Another aspect of common courtesy is knowing when to shut up. Bots should be aware of when the user would like to suspend or delay the conversation. Supporting commands like "pause," "stop," or "dnd," or just delayed responses, is recommended in many use cases (Figure 8-33).

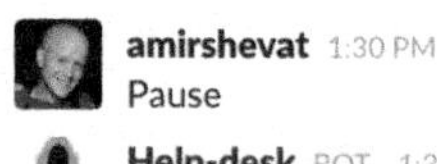

amirshevat 1:30 PM
Pause
Help-desk BOT 1:30 PM
Roger that! I will wait here until you get back to work on this support ticket.

FIGURE 8-33.
Handling a pause command

Other aspects of common courtesy might be giving the user enough time to perform an action, providing sensitive information privately (we will talk about that in depth in the next section), and being empathetic to the user's needs and pains. All of this will result in a nicer and more humane bot that is easier and more pleasant to work with.

Team Versus Private Interactions

Having a conversation as a team might be a completely different experience to a private conversation. Bots working in a team context need to know how to work with multiple inputs from multiple team members. Figure 8-34 shows an example.

LunchBot BOT 5:01 PM
who is ready to order lunch? We have Pizza and Burgers today!

FIGURE 8-34.
Accepting input from multiple users

The bot in this use case needs to know how to work with multiple users in the same channel or environment. The bot needs to acknowledge input by user and communicate to different members in the channel.

In team environments, the bot can participate in different types of conversations. Let's take Slack for example. The options include:

1. *Public channel*—A channel that is open to all members of the team. The #general channel is an example of such a channel.
2. *Private channel*—A channel with limited access. Only invited team members can participate in this channel. Members can be added to or removed from the channel at any time.
3. *Multi-party direct message (MPDM)*—A private conversation that is happening between a limited set of users. Users cannot be added to or removed from this conversation after it is initiated.
4. *Direct message (DM)*—A private conversation between two users. The bot can be one of these users.

Each of these interaction modes can be used by a bot in a conversation. In a channel conversation (or MPDM), the bot needs to be added to the channel in order to send messages to and receive messages from users in that channel.

When the bot receives a message from a user in a team environment such as Slack, the bot will receive meta information such as the team ID, the user ID, and the channel ID. Bots can also list the users in a given team or channel and build their conversation accordingly. Bots can even create a group conversation by creating a channel or an MPDM, adding team members to it, and initiating a conversation there.

CHOOSING THE RIGHT INTERACTION MODE

When designing a conversation in a team context, you have the flexibility to start a conversation in one mode and move to another mode, based on the privacy needs and process requirements.

Here is how a hiring bot conversation can move from a private conversation to other modes of interaction:

1. Private hiring channel, #hire-requests:

   Bob (hiring manager): @hiring-bot please start a hiring process for a new frontend developer - Lili Cheng lili@gmail.com.

   Hiring-bot: Sure Bob! Starting a hiring process now. @Steph - please confirm we have the headcount.

   Steph (HR manager): Confirmed

2. Direct message (Bob and bot):

   Hiring-bot: Hi Bob! Your candidate, Lili Cheng, has passed 3 interviews successfully!

   Bob: Fantastic! Please proceed to giving her an offer (L7 Eng)

3. MPDM (Steph, Bob, bot):

   Hiring-bot: I would like to confirm the compensation details for candidate Lili Cheng. Here are the package details: ...

   Bob: Sounds good to me

   Steph: Yes, sounds good.

4. Public new hire channel:

   Hiring-bot: Hi everyone, please welcome to the team our latest new hire, Lili Cheng!

   Mike Brevoort: Yaay!

   Lauren Kunze: Welcome to the team!

   ...

   Lili Cheng: Thanks everyone!

As you can see, the bot manages the hiring process, moving from private chat, to multi-party direct message, to public announcement. If you are building for a team, you will need to inspect your communication and see what is the best mode of interaction.

Things to think about when choosing an interaction mode:

Data sensitivity
: You might want to share some things in private with some users (for example, the compensation package in the case of our hiring bot).

Private/personal/restricted interactions
: A direct 1:1 bot is great for personal task management such as to-do bots and personal assistants.

Team dynamics
: Will you need to loop people into the conversation? Is there an acceleration path you need to follow? Remember that members can be added to private channels but not to direct messages.

Team/bot culture
: Sometimes peer recognition is done in private, but some teams love to put it in a public channel.

Compliance
: Direct messages are not always accessible to admins and regulation apps.

Some interactions can be done in private mode more easily because the bot needs the user @mentions to direct messages to a specific person in a channel. Direct messages do not require an @mention.

USING @MENTIONS

In some messaging platforms the bot can use an @mention to direct a message to a specific user. This denotes the message visually as being intended for a particular user, and usually creates a notification/alert for the mentioned user.

Here are a few examples of @mentions:

> ***Bot***: @amir - would you like to set this meeting up?
>
> ***Bot***: Setting the all hands meeting for April 4th. CC: @amir @jassim
>
> ***Bot***: As agreed by @ceci and @amir, we will hold the meeting this Friday!

All these @mentions create notifications for the mentioned users. Remember that @mentions are a way to escalate a message. Users might read a message even when they are not @mentioned, but using the @mention will send them a notification that will improve the chances of them seeing the message.

Some @mentions are special. Some platforms, like Slack, support @here, @channel, and @group-name; all of these are a way to notify and escalate the conversation to multiple members of the team.

Bots can also be @mentioned—in Kik, a bot will receive a message from a user using an @mention of the bot's name even if the user has not installed the bot. In Slack, developers can filter messages coming as direct mentions and run code whenever that happens. Using the @mention in Kik is a good way to improve bot discovery—when two users talk about a bot and the bot is getting a notification of the conversation, it could be a way to initiate bot interaction.

Creating channels is also a great way to facilitate processes. In Slack, for example, every feature is represented by a #feat-<feature-name> and launches are discussed in a #launch-<feature-name>. In addition to creating channels, bots can archive channels when a process is done. Creating a channel, inviting the relevant members, and archiving the channel at the end of the process can be a great way to utilize the team and channel infrastructure to implement business processes.

Remember that channels are a limited resource in some platforms and creating too many channels can impair the user experience, so if you need to create a lot of channels on the fly you might want to consider an alternative.

TRAINING AND ONBOARDING

We discussed onboarding earlier, but it is important to stress that onboarding is very different in a private conversation with one person and in a team environment, where a bot is installed by one member and then used by multiple members.

Use the onboarding of the bot to train and educate users about their role in the process that the bot is facilitating. You can also listen to the event of a user being added to a team or a channel and trigger a personal training session.

Here are a few examples:

> ***Bot***: Hello Team! @don just installed me :) I am a coffee ordering bot and you can use me starting today to order coffee at the expense of the company! Hurray! DM me anytime with "coffee!"

> ***Bot***: <DM> I see you have joined the legal channel. I am the legal bot and will help you review legal documents. You just need to upload contracts to the #legal channel and I will send you comments about them.

KNOWING WHEN TO SHUT UP IN A TEAM CONVERSATION

One of the important things to remember about a bot in a team conversation is that most of the communication is not directed at the bot. The bot might "hear" a lot of noise that is not relevant to the process it needs to run.

Bot builders need to learn to filter the signal from the noise, and only reply to messages that are addressed to the bot or that are a part of the conversation it is having. Using rich interaction can mitigate some of this confusion; we will discuss this in the next chapter.

Error Handling

To err is human; to fix the error and get the conversation going again is your responsibility.

Most bot developers are astonished when they look at their logs and see what humans are saying to their bots. Conversational divergence is very common, and error is just a part of it; many users provide faulty information to a bot just to "play" with it and see how smart it is.

Looking at logs, developers report human inputs including:

> *User*: dgdg dffdgd
>
> *User*: What are you wearing today?
>
> *User*: Are you bot or human?

Irrelevant conversation is very common in bot interactions. Figure 8-35 is a real-life example of a user conversation with the Poncho weather bot.

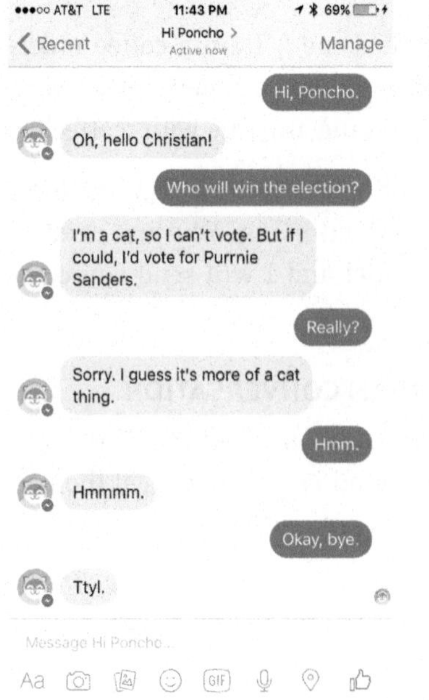

FIGURE 8-35. Poncho getting inputs that are off-topic

This sort of thing happens often and can occur at any point in the conversation.

Examples of divergence in conversations can be very subtle, and this may be done with positive intent. Let's go back to our broken conversation example from earlier in this chapter:

> *User*: Hello @coffeebot
>
> ***Coffee-bot***: What would you like today? We have regular coffee and espresso.
>
> *User*: Wait, no! I want a cappuccino!

There are various ways you can proceed from here.

COURSE CORRECTION

Course correction relies on the bot's ability to pull the user back into the happy flow of the conversation.

Here is an example of getting the conversation back online and restricting the user to the happy flow:

> ...
>
> *User*: Wait, no! I want a cappuccino!
>
> ***Coffee-bot***: We do not have the requested coffee today - we have regular coffee and espresso. What would you like today?
>
> *User*: OK, let's go with the regular coffee.

There is another way to handle a request from a user that cannot be fulfilled at the moment, and that is to collect that information and use it to grow your product.

Here is an example:

> ...
>
> *User*: Wait, no! I want a cappuccino!
>
> ***Bot***: We currently do not serve cappuccino (you can order regular coffee or espresso), but I will also notify you when we start to serve cappuccino.

The bot can also have a "special coffee" request flow built into it, potentially involving human intervention, but that is a complexity we will avoid at this point and talk about in the next section.

Josh Barkin, cofounder and CEO of Wordhop, a CRM toolkit built for the messaging era, had an interesting insight about course correction:

> If the communication breakdown with users occurs in the middle of a conversation, and you're able to maintain context, you can serve up two messages in succession. The first is the fallback message to indicate misunderstanding, and the second is a reminder of the context. In this sense, you're able to move the focus back to your goals and not let the misunderstanding become a distraction in your user experience [see Figure 8-36].

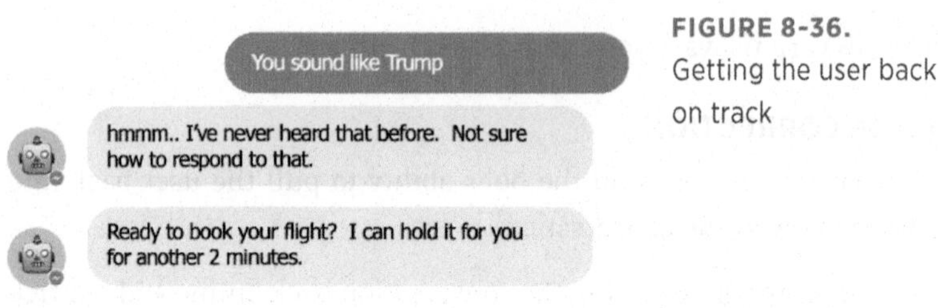
You sound like Trump
hmmm.. I've never heard that before. Not sure how to respond to that.
Ready to book your flight? I can hold it for you for another 2 minutes.

FIGURE 8-36. Getting the user back on track

> Another redirection strategy is to ask the user if they would like the bot to escalate the request and if they concur, you're able to funnel the user into another response channel and minimize the risk of losing them.

While this seems trivial, course correction is still quite uncommon in bot conversations. We still see a lot of bots just doing this:

> ...
>
> *User*: Wait, no! I want a cappuccino!
>
> ***Coffee-bot***: I do not understand your request.

This is what we call "dead end"—a savvy user will try to reignite the conversation, but most users will get frustrated and abandon the conversation. A dead end is the worst possible outcome, other than outputting an error with debugging information (I have seen that as well). Even if you do not want to course correct, send the user to a viable flow in the conversation.

HUMAN INTERVENTION

Deferring to a human supervisor is a viable and common solution to error handling with bots. This could be transparent to the user, where a human supervisor intercepts the conversation and takes over, or the bot can state that it is deferring to a human to get the problem sorted:

> ...
>
> *User*: Wait, no! I want a cappuccino!
>
> ***Coffee-bot***: Let me connect you with my human supervisor.

The human can then follow up on the conversation and let the user down with empathy, or go to the coffee shop and get them a cappuccino as an exception.

It is important to note that users interacting with bots might develop expectations of immediate response, which might not be available when a human follows up. So, setting expectations about when the user will hear back from the human might be a good idea:

> ***Coffee-bot***: Let me connect you with my human supervisor. This might take a few minutes.

You can build a process of course correction, followed by human intervention if that fails. In some instances, support like this is an expected pattern—the bot serves as a first line of defense and the human supervisor steps in to provide assistance in cases where the bot fails.

RESTARTING THE CONVERSATION

While it can be annoying, restarting the conversation might be a viable choice when you cannot have a human supervisor who can step in or cannot effect a course correction, navigating the user back to the last point in the conversation. Restarting the conversation should be avoided, but is still better than a dead end in many use cases.

REDIRECTING TO ANOTHER BOT

This is not a common practice, but it can be a delightful one. This insight was shared with me by Dr. Barbara Ondrisek, founder of Mica, the Hipster Cat Bot, a bot that helps you discover hip places around the world:

We realized we could not create an omnipotent bot that answered all the questions in the world. Even Siri and Alexa have their limitations. But what we did was to redirect people to other bots! When users ask for the weather, Mica says the following [Figure 8-37]:

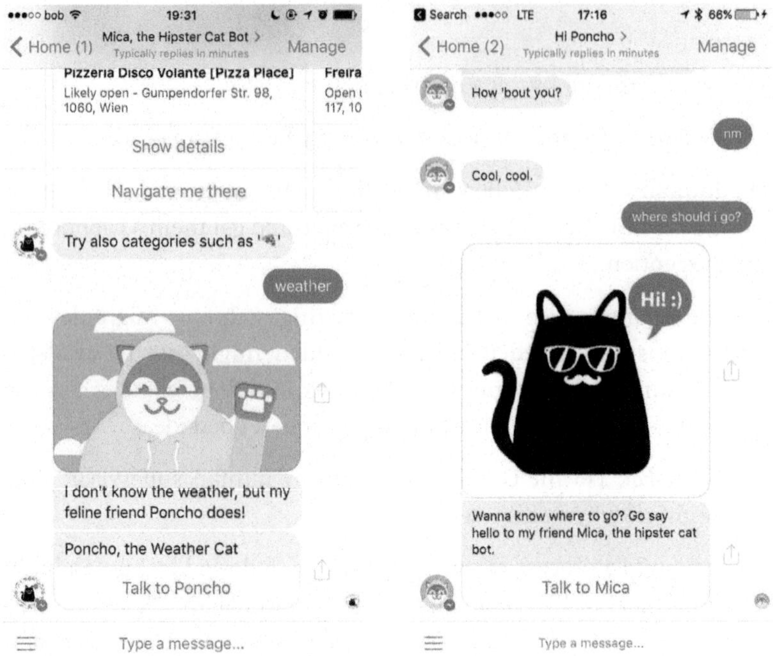

FIGURE 8-37.
Bots recommending other bots

Mica recommends Poncho, the weather cat bot! And when users ask Poncho where to go, he responds with a link to Mica. In the future bots might also talk to each other!

By connecting the user with another bot that can address the user's intent, Barbara is able to resolve an error, delight the user, and generate traffic to a fellow cat bot. In the future there might also be a potential revenue stream that can come from bots sending users to each other.

KEEPING IT CONSISTENT

It's important to note that whatever strategy you apply when handling errors, your bot must keep a consistent persona throughout the conversation. A delightful banking bot that is "always happy to help" cannot start spitting "error 524, invalid input"—that would create dissonance and will antagonize the user.

Oren Jacob, the CEO of PullString, a conversational tool we will review later, had a wonderful quote: "Darth Vader's true personality will not show when you ask him about the Death Star, it will show when you ask him about the price of tomatoes." The character of your bot will shine in edge cases where the bot does not know how to handle the user's requests.

LEARNING FROM YOUR BOT'S MISTAKES

Conversation design should be done with a growth mindset. Designing a conversation is an ongoing process of learning from your bot's mistakes. Looking at all the intents your bot failed to understand, all the failed conversations, and all the requests that your bot could not fulfill, is a wonderful learning experience. Most bot designers collect these conversations and sort them by how often they happen. If the bot constantly misses a specific intent or offers the intent too often, it might be time to fine-tune the text-to-intent mapping. If the users keep asking the bot to do something it can't, it might be a good opportunity to build a process to fulfill that need.

[KEY TAKEAWAY]

Designing a conversation is an ongoing process of learning from your bot's mistakes.

A secondary iteration is capturing frustration words that indicate that the user is not happy with the conversation or the service and analyzing them.

This is an ongoing exercise—your bot should always be "growing and improving" and the bot's design should be continually optimized.

Help and Feedback

In many bots, the help and feedback sections of the conversational interface are overlooked. Much like onboarding, offering help is critical to the bot's success, and feedback provides important information to you that you can use in order to improve your bot's design.

PROVIDING HELP

Help should always be available to the user—if a user at any point in the conversation says "Help" or "Help me" or any variant of this, the bot should move to a help mode. Help can be as simple as repeating the section in the onboarding script that teaches the user how to use the bot. Figure 8-38 shows how I implemented it in my Wordsbot.

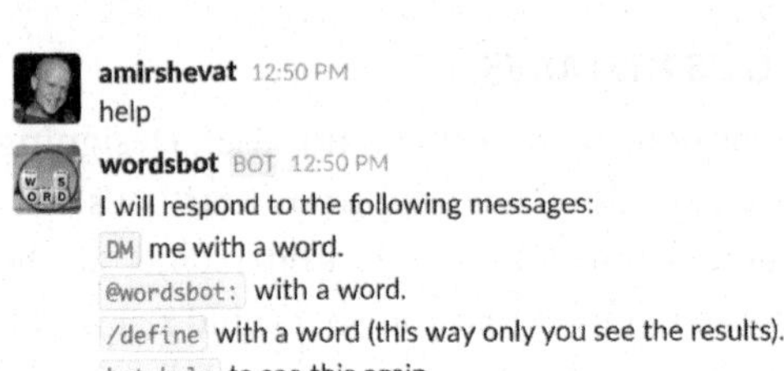

amirshevat 12:50 PM
help
wordsbot BOT 12:50 PM
I will respond to the following messages:
DM me with a word.
@wordsbot: with a word.
/define with a word (this way only you see the results).
bot help to see this again.

FIGURE 8-38. WordsBot responding to a request for help

This is a pretty simple and effective method with task-led bots. The bot reminds the user how to use the bot to achieve the required task. In a more complex, multitask bot such as Google Assistant, you might want to point out the top-level navigation tasks (Figure 8-39).

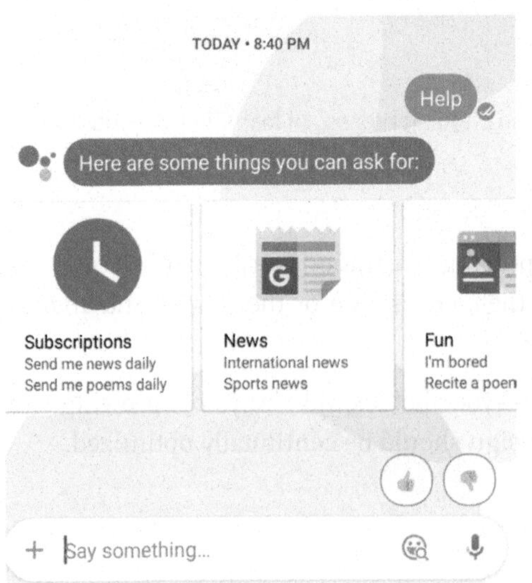

FIGURE 8-39. In Google Assistant, requesting help takes you back to the top-level navigation

An advanced version might be to provide contextual help, relevant to the point of the conversation the user is at with the bot.

For example, in our coffee bot it could work like this:

> ***Coffee-bot***: What would you like today? We have regular coffee and espresso.
>
> *User*: Help <or Help me choose>
>
> ***Coffee-bot***: Help with choosing a coffee: Espresso is a small, mid-day drink popular in Italy. Recommended for a quick energy boost.
>
> ***Coffee-bot***: Regular coffee comes in 3 sizes—large, medium, small—and is popular in the USA. Recommended for sit and relax times.
>
> ***Coffee-bot***: For general help please type help again.

In this advanced use case you need to build help flows for different steps of the conversation. These help flows can drive conversions, guiding users through the conversation funnel—you might even prime the user to make the "right" choice with a creative help flow.

Help should be available at any point, but in a team context help text provided for one user might be seen as spammy by other members of the team, and it might be better offered in a private or ephemeral (visible only to the user) way. Some platforms let you send private messages to the user, either in the channel or in a direct message. Unless the help text is likely to be useful in a team context, it is best to provide the help in a private environment.

Analyzing when a user is requesting "Help" can contribute to the improvement of your bot. If you notice that a lot of users get stuck and ask for help in a particular place in the conversation, it might be wise to rework the script or even restructure the conversation. Requests for help might indicate a broken part of the conversation, most commonly around initial bot engagement.

SOLICITING FEEDBACK

Feedback is a way for users to provide you with information about their experience with the bot. Google Assistant enables this throughout the interaction (Figure 8-40).

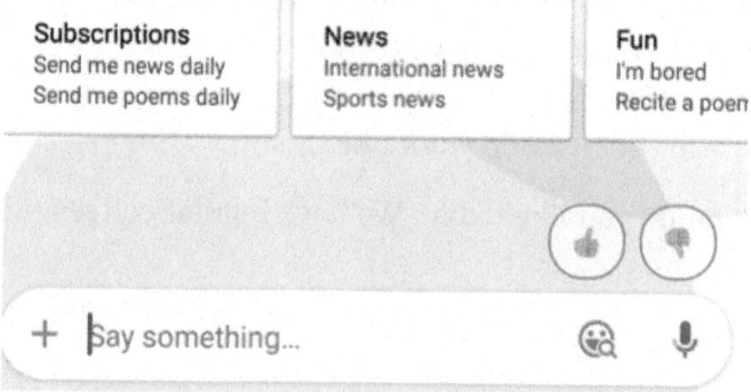

FIGURE 8-40.
Google Assistant invites user feedback throughout the experience

This might look to you a little too intrusive as a way to ask for feedback, but this is an early beta version of the Google Assistant bot, and the team is really looking for user input.

There are other ways to gather user feedback in later stages of your bot's conversation:

Support the "feedback" command
: This should be a best practice for all bots. Whenever a user says "feedback," start a feedback conversation.

End a conversation with a request for feedback
: Especially when the conversation is task-based, ask for the user's feedback at the end.

Capture keywords
: In many cases you can get implicit user feedback by looking for certain words. Users tend to say "thank you" and even "I love you" when a task is well done. You can guess the words your users might say when they get lost or are having problems with the bot's functionality. Capturing these keywords and analyzing them can give you a lot of insightful information.

Closing Thoughts

In this chapter, we covered different aspects of text-based conversation—we discussed the importance and role of onboarding, then reviewed topic-led conversations, task-led conversations, and the key differences between them.

Next, we covered some best practices when thinking about conversations in both a team context and a private context. We reviewed ways to optimize the conversation and handle errors and feedback, and saw a few tricks that can prime the user to provide the inputs we need in the format we need them. This was a lot of theory to digest—we will put this theory into practice in Chapters 16 and 17.

In the next chapter we will review rich interactions and controls that can help us create better conversations.

[9]

Rich Interactions

Use a picture. It's worth a thousand words.

—TESS FLANDERS

RICH INTERACTIONS AND CONTROLS are a great way to simplify, optimize, direct, enrich, and sometimes just replace text-based conversations. These are also the most volatile aspects of each bot platform. Most platforms support multiple types of rich interactions and controls, but each platform implements these in a different way (similarly to how in the mobile world, a button will look, and even act, differently on Android and iOS). Because of all of the permutations and ongoing changes, it would be extremely tedious and ineffective to cover all the nuances of each platform. We will review different implementations and show differences between platforms, but we'll focus on the most common chat platforms and use cases.

Fine-tuning the balance of rich interaction and text-based interaction is a matter of understanding your use case. In a task-led conversation you might want to over-index on rich controls, in order to work around lengthy conversations. In a topic-led discussion, you might want to over-index on natural conversation, and let the user enjoy the chitchat and lengthier discourse.

[KEY TAKEAWAY]

Rich interaction and controls are a great way to simplify, optimize, direct, enrich, and sometimes just replace text-based conversations. In a task-led conversation you might want to over-index on rich controls, in order to work around lengthy conversations.

Files

Almost all platforms support multiple file types that can be added to the conversation—both users and bots can add files as part of their interaction, enriching the conversation as well as facilitating business processes.

The LawGeex bot is a legal bot that helps you evaluate contracts. Describing the contract in a conversation would be impossible, and pasting it as text in a chat would be cumbersome as contracts are usually kept in *.doc* files. As an alternative, the LawGeex bot asks the user to upload the file and starts the review process once the file is uploaded (Figure 9-1).

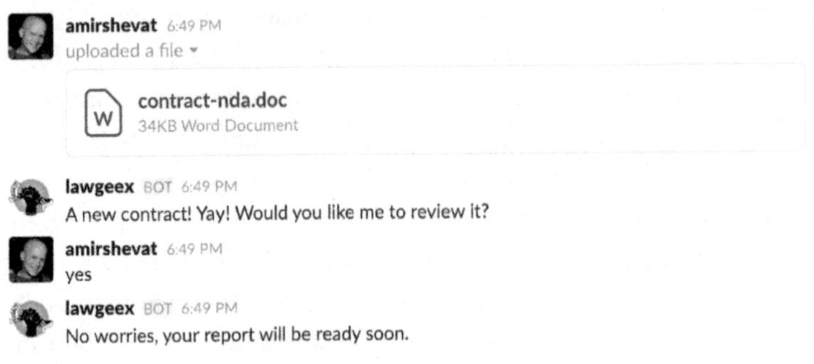

amirshevat 6:49 PM
uploaded a file
contract-nda.doc
34KB Word Document
lawgeex BOT 6:49 PM
A new contract! Yay! Would you like me to review it?
amirshevat 6:49 PM
yes
lawgeex BOT 6:49 PM
No worries, your report will be ready soon.

FIGURE 9-1.
LawGeex allows users to upload contracts as .doc files

There are many other file formats your bot can support, and we will dive into some of the more common ones at length in this chapter. But as a general best practice, think about how you can use files as part of your flow. Even taking the simple example of the coffee bot from the previous chapter, you are mandated by law to provide a receipt at the end of a business transaction. Posting a receipt after checkout as an image or a *.pdf* file would be a great way to fulfill that requirement. This is also useful for your users because they can always go back and search for their coffee receipts.

When a user uploads a file in a format that your bot does not support, the bot should provide an error flow that will help the user understand what to do. For example, if the user uploads a *.docx* file while your bot only supports *.doc* files, the bot should guide the user to repost the file in the appropriate format. Even if your bot does not support any file input, it should expect to get files and provide an appropriate error/help flow.

WHEN TO USE FILES IN A CONVERSATION

Not every conversation needs to involve files. Use files in a conversation when the work being done is on the file itself. Examples might be Word docs in enterprise productivity use cases, or PDF files of detailed reports or read-only documents such as receipts or purchase orders). Think of use cases where it is easier for the user to upload a file to the chat instead of typing in the information manually, as in a case of expense reports. Remember, be ready to receive a file from a user and handle it with grace, even if your use case does not support files.

Audio

Audio is an important aspect in some bots. In many households, Alexa serves as a music player. "Alexa, play X" is a commonly used skill of this Amazon bot.

Audio files are also useful in use cases where you want to post recordings in a conversation, in order to store and collaborate on them. An example of that would be a bot that helps you to initiate a call and stores a recording of the call in the chat client.

Audio serves for some bots as the primary input mechanism for the user. From Siri to Google Home, the primary interface of most super bots is a vocal user interface (VUI). As we are focusing on chat platforms in this book, we will talk only briefly about this type of interface, later in the book.

Note that audio/voice might not be an effective medium in use cases where users do not feel comfortable playing the bot's output or voicing their intent out loud. This type of engagement is usually optimal for hands-free, private (not in public) interaction.

Videos

Videos are a great way to relay information, entertain, and provide a rich experience that is very different from a text-based experience. Videos are an engaging format for many users, and especially for younger audiences. Many bots on Kik use videos to enrich the interaction, from providing style review videos to funny video games. Some video services have actually created video bots on this platform (Figure 9-2).

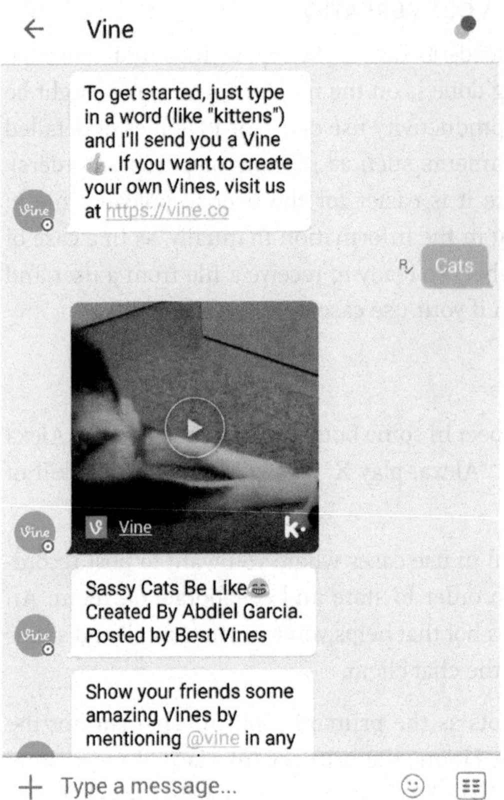

FIGURE 9-2. The Vine video bot on Kik

In other use cases, you can use videos as part of your onboarding script to help users understand how to use the bot. Bots can also pull videos from other services and post links into a conversation—think of a news bot that posts links to videos of news events as they happen. We will discuss more about posting links in a later section of this chapter, but some links from trusted video providers can unfurl (create a playable preview) and show inline in a chat client.

In the example in Figure 9-3, I've asked the Help-desk bot to provide help with restarting an iPhone. Providing a link to a YouTube video that explains how to do this is a lot more useful than describing it in text. Notice that the bot does not upload the video; just posting the link in the conversation makes it available to watch embedded inline in the Slack client.

Help-desk BOT 8:56 PM
Here is a video that might help -
https://www.youtube.com/watch?v=j0Tlmsx2ETM
YouTube How to Smartphone
iPhone 6 / iPhone 6 Plus - How to Soft Reset. (Clears minor malfunctions)

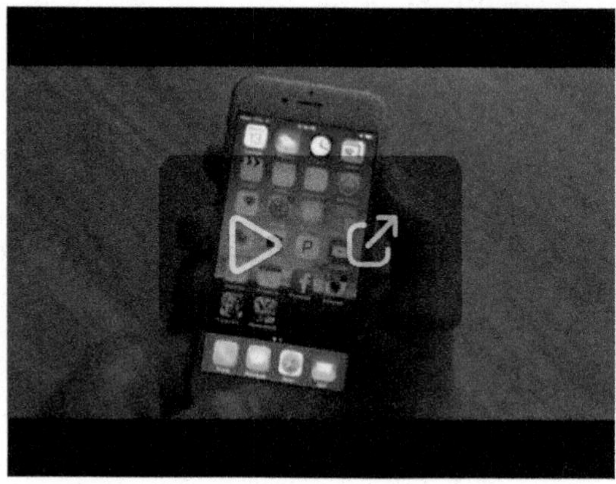

FIGURE 9-3.
A short video can be worth a thousand words

WHEN TO USE VIDEOS IN A CONVERSATION

Sometimes it's appropriate to pull preexisting videos into the context of the conversation. Find use cases where videos provide a more compelling engagement, such as product reviews. Explore using videos to educate the users about your bot, its usefulness, and ways to work with it (you can use videos or animated GIFs, as discussed next). Remember that users can upload videos into a conversation with the bot as well.

Images

Images are one of the most common rich interactions across all platforms, and probably the most visually consistent. Images are useful as some information is much easier to relay in an image than in words.

Figure 9-4 shows how the Kip bot uses images when presenting shopping search results on Slack.

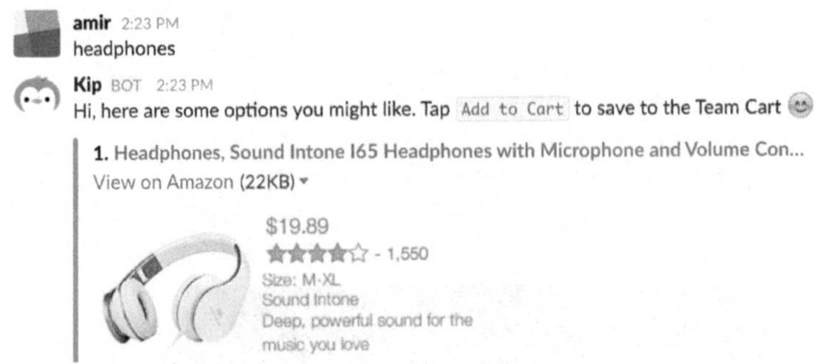

FIGURE 9-4.
The Kip shopping bot displaying search results on Slack

Figure 9-5 is an example from the same bot on Facebook Messenger.

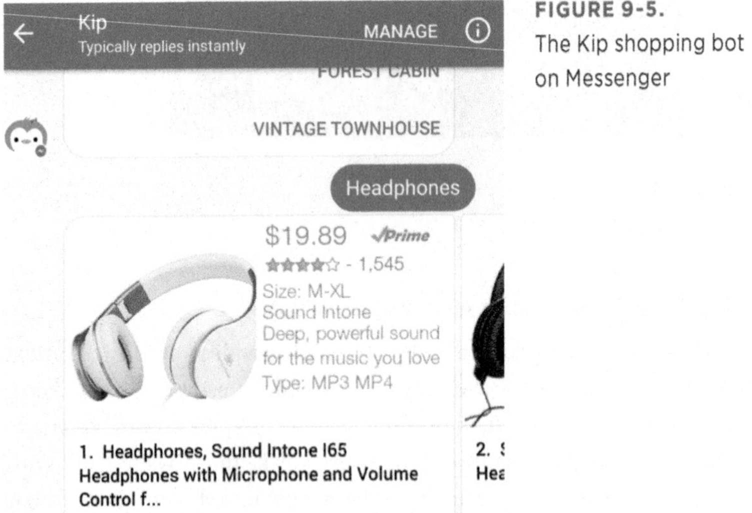

FIGURE 9-5.
The Kip shopping bot on Messenger

As you might have noticed, Kip's designers compacted a lot of information into the image: a photo of the headphones, the price, the rating, and a text description. The team overcame some of the platforms' limitations (for example, there are no star controls built into any of the current platforms) by adding elements to the image on the server side, and then serving the image in the conversation.

As you can see, the image part of the message is consistent—the size and the way multiple images are presented might be different across platforms, but other than that the controls behave the same. The image provides a lot of information that it would have been hard to convey in a text-based conversation, and even if it were described accurately, the user might have a hard time visualizing the product without the image.

Images can also help users visualize data more clearly and easily than is possible with text. Figure 9-6 shows an example of how Statsbot provides Google Analytics information in Slack.

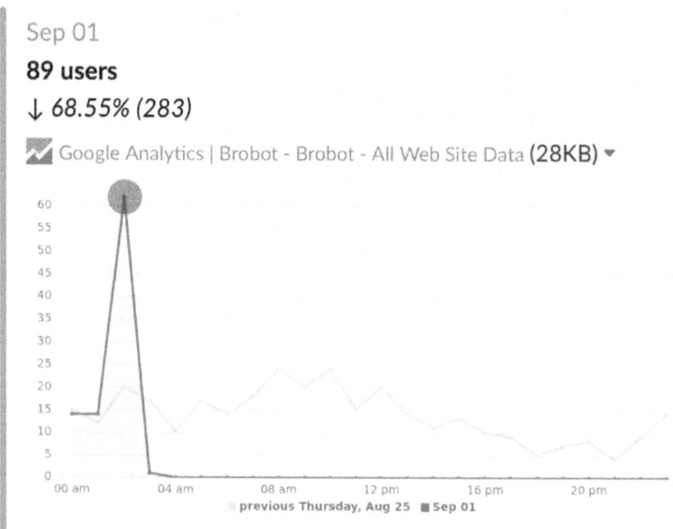

FIGURE 9-6.
Statsbot's charts

Relaying this type of information is almost impossible to do effectively with just text. Statsbot uses text for the verbal analysis, combined with an image taken from Google Analytics.

Images can also be sent from the user's side. Google Assistant does a great job of interpreting the image's content and offering the user actions associated with that image (Figure 9-7).

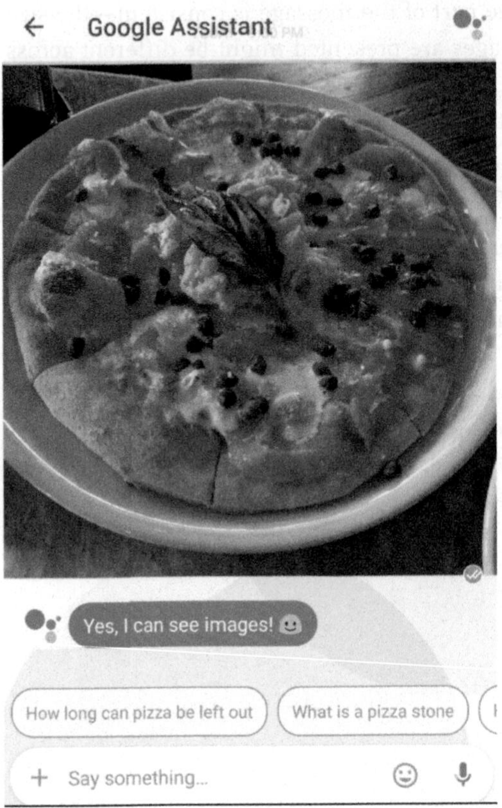

FIGURE 9-7.
Google Assistant recognizes when the user posts a picture of pizza and suggests related topics to explore

In this example, the user uploaded a photo of a pizza. Google Assistant recognizes the image content and suggests relevant actions, such as searching for "how long can pizza be left out?" Users can upload photos of things they cannot describe in words, and the bot can interact with the content without needing a textual description.

Another form of image is the animated GIF format—this is a short, animated "video" image that can be displayed in most places you would be able to display a static image. We will see an example of how to use an animated GIF during onboarding in the next chapter. GIFs are a great way to display short, audioless video content, as they are lightweight and widely supported by most platforms.

Images are also effective at improving engagement. The Poncho weather bot adds a funny image to every daily notification. Users are delighted by this feature; some report that that image is sometimes more interesting to them than the weather forecast content itself.

WHEN TO USE IMAGES IN A CONVERSATION

Use images often to enrich the conversation. You can include images inline to delight the user or entice them to take action, use images to showcase products in ecommerce use cases, and include graphical reports and charts to help users visualize information. You can also use animated images (GIFs) to explain processes—when onboarding users, for example—or to share fun and engaging content as part of the conversation.

Buttons

Buttons are arguably the most useful rich control available today for conversational interfaces. Buttons can be a great way to guide the conversation, frame the interaction, limit the user to a set of answers, provide the user with a set of canned responses, enable navigation in an app-like manner, and more.

> **[KEY TAKEAWAY]**
>
> Buttons can be a great way to guide the conversation, frame the interaction, or limit the user to a set of options.

Buttons are implemented in different ways on different platforms.

BUTTONS IN SLACK

Buttons in Slack can be added to a message sent by the bot, in a container technically called a message attachment (see Figure 9-8).

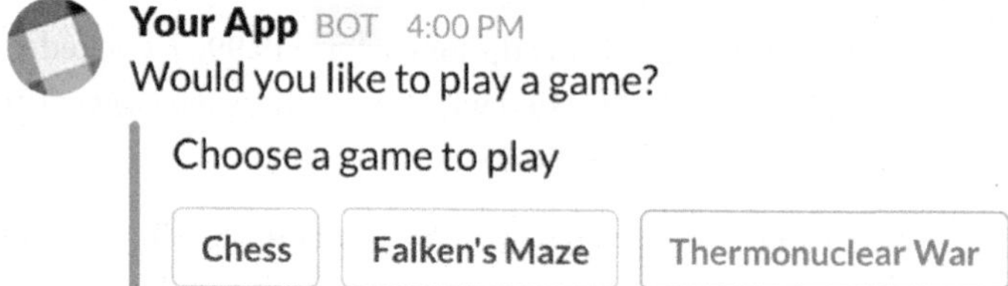
Your App BOT 4:00 PM
Would you like to play a game?
Choose a game to play
Chess
Falken's Maze
Thermonuclear War

FIGURE 9-8.
Message buttons in Slack

Buttons can have text (including emojis) and a style (danger, primary, or default, following the web styles). The recommendation is not to use the emoji in the button itself, and not to overplay the buttons' styling—only one button should be either danger or primary, and the rest should be default.

Once a button is clicked, one of three things can happen:

1. The bot can *post a new message*. For example, the bot can continue the dialog based on the user's button click. Figure 9-9 is an example from Kip.

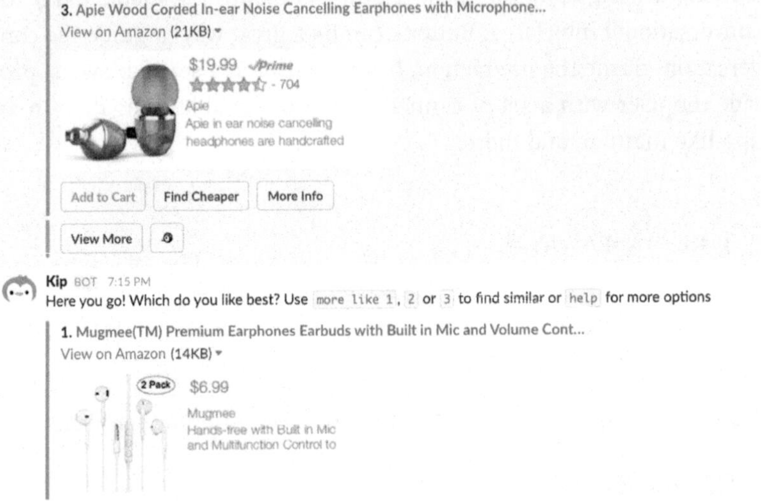

FIGURE 9-9.
Kip decides how to continue the conversation based on which button the user clicks

   The bot initially offered me headphones for $19.99; I clicked on "Find Cheaper" and the bot replied with a new message suggesting some cheaper headphones for $6.99.

2. The bot can *replace the original message*, including the buttons. This is very powerful because it enables the bot to provide app-like navigation within the same message context. Figure 9-10 shows how a DevOps bot called Beep Boop uses buttons for navigation.

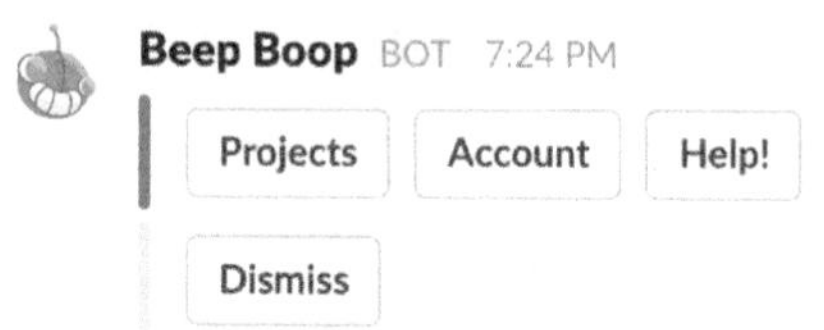

FIGURE 9-10.
Beep Boop uses buttons for navigation

The bot provides the user with simple navigation buttons—let's click on the Account button (Figure 9-11).

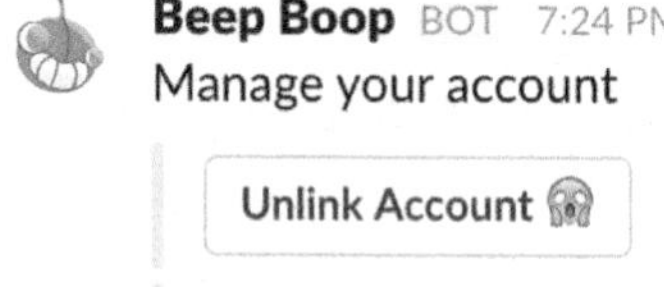

FIGURE 9-11.
The result of clicking on the "Account" button

Beep Boop replaces the original message and provides us with an option to unlink the account, go back to the main menu, or abort the conversation altogether by clicking on Dismiss.

Another option is to remove the buttons altogether and change the message to be the result of the task. In Figure 9-12 the user is asked to approve or reject a request to make an offer to a hiring candidate.

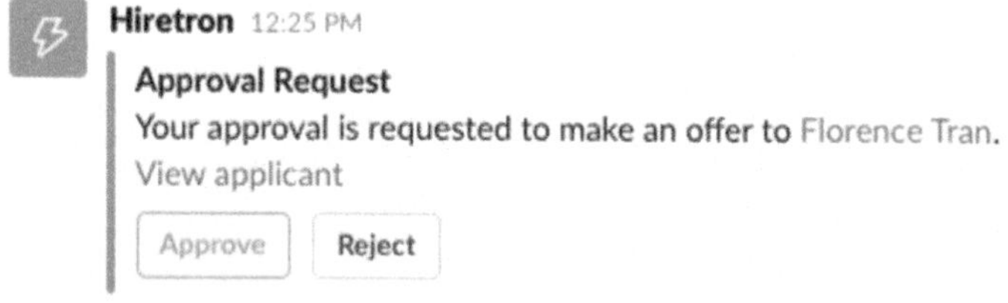

Hiretron 12:25 PM
Approval Request
Your approval is requested to make an offer to Florence Tran.
View applicant
Approve
Reject

FIGURE 9-12.
Buttons can be used to ask a user to approve or reject an action

After the "Approve" button is clicked, the message turns into Figure 9-13.

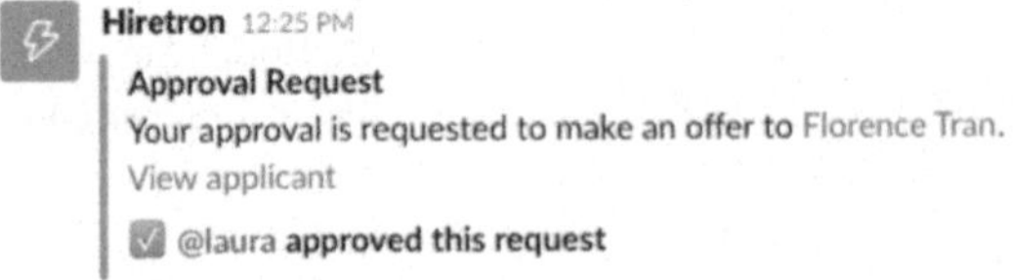

FIGURE 9-13.
After approving the action, the message updates

As this type of approval task does not require the buttons once the request has been approved, the message changes to show what action was performed, and by whom.

Another interesting use of buttons to change the message is using them to filter the content. For example, a flight bot might provide the user with three flight results, and filter buttons that say "Night Time Flights" and "Direct Flights." Clicking on one of these buttons would replace the content of the message accordingly. Similarly, pagination can be implemented using buttons—the flight bot might provide the user with three options and "Next" and "Previous" buttons for pagination.

Figure 9-14 is an example of how the Kip team implemented counters and checkboxes using buttons: they added an unchecked button and replaced it with a checked button once the button was clicked.

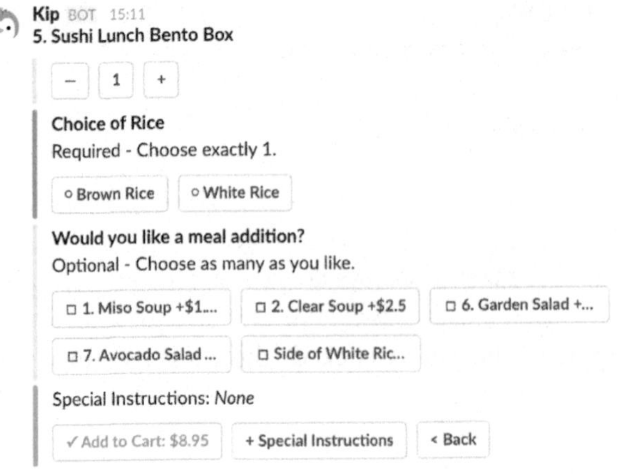

FIGURE 9-14.
Updating the message to show a "button clicked" state

As you can see, the Kip bot simulates a lot of different controls using the ability to replace the button once it is clicked. The team really hacked a brand new set of controls using only buttons and inline replacement.

3. Lastly, a bot can *prompt the user with a confirmation window* once a button is pressed. Figure 9-15 shows what this looks like.

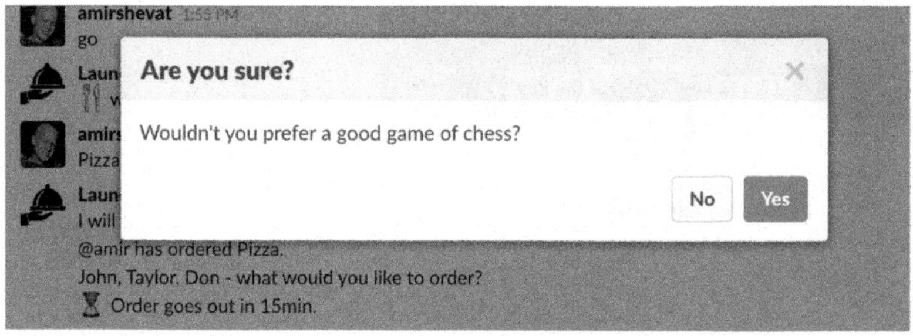
amirshevat
go
Are you sure?
Wouldn't you prefer a good game of chess?
No
Yes
@amir has ordered Pizza.
John, Taylor, Don - what would you like to order?
Order goes out in 15min.

FIGURE 9-15.
Presenting a confirmation window in response to a button press

Clicking on one of the buttons in this confirmation window can trigger the next step in the flow. This process is great when additional confirmation is part of your workflow.

Slack's approach to buttons, being a business-facing platform, is focused on business workflows, facilitating getting work done and use cases that involve requests, approvals, assignment of tasks, data analysis, requisitions, sales, and so forth.

BUTTONS IN FACEBOOK MESSENGER

Facebook Messenger, being a consumer-facing platform, has taken a slightly different approach to buttons. Buttons in Messenger focus on common consumer flows, navigational controls, and input-capture controls. Here are a few interesting use cases:

1. You can *add an onboarding Get Started button* to the bot's welcome screen. The welcome screen is the first thing people see when they start a new conversation with your bot. This is a great way to prime the user to make an initial engagement with the bot and to set the context for the bot.

As you can see in Figure 9-16, a lot of additional information about the bot is included on this screen. (We will review different aspects of discovery in Chapter 11.)

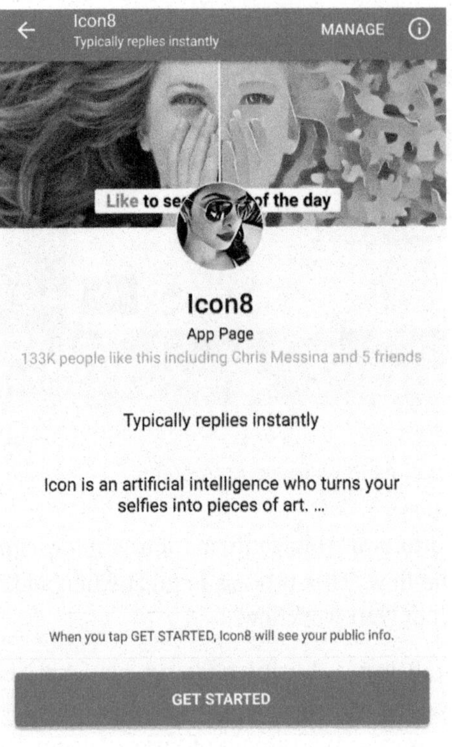

FIGURE 9-16.
The welcome screen for the Icon8 bot

2. A bot can *add buttons in templates.* We will discuss templates later, but you can think of them as a well-structured composite of UI elements.

   Buttons in templates can do several things:

   - Open a new, separate web page (*URL buttons*)
   - Generate a callback to your backend server that can trigger the bot to take action or converse with the user (*Postback buttons*)
   - Trigger a native sharing process, useful for enabling the user to share content from the conversation with the bot (*Share buttons*)
   - Start a checkout process for items the users can buy as part of the conversation (*Buy buttons*)

- Initiate a call on a mobile device (*Call buttons*)
- Link/unlink the user in the conversation with a third-party account system (*Log In and Log Out buttons*)

Figure 9-17 shows an example of template buttons in theScore bot.

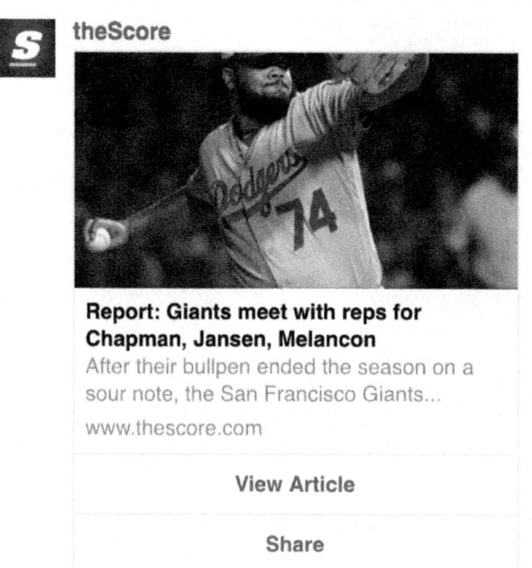

FIGURE 9-17.
URL and Share template buttons in theScore bot for Facebook Messenger

The bot sent the user an update about a sporting event and added two buttons to the template: a "View Article" button that opens the article in the browser and a "Share" button that opens a share flow in Facebook.

CANNED RESPONSES IN FACEBOOK MESSENGER AND KIK

In Facebook Messenger, a bot can specify canned responses called Quick Replies. When a user clicks on a Quick Reply, the selected response is sent to the bot. The Quick Replies prime the user to choose one of the canned responses. It is a good way to steer the user to the right path in the conversational flow. Figure 9-18 shows how you can combine template buttons and Quick Replies in the same flow.

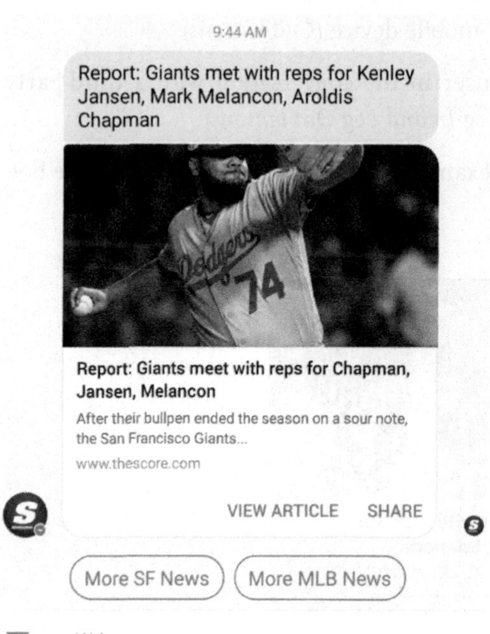

FIGURE 9-18. theScore bot offering both template buttons and Quick Replies

Note that the canned responses show up at the bottom of the chat window, just above the input control, hinting to the user to pick one of the Quick Replies. It's also important to note that canned responses suggest to the user what to respond, but the user can still ignore the canned responses and type in free text—so remember that the bot needs to handle both Quick Replies and free text.

One interesting Quick Reply provided by Facebook is the "share location" one. When a user clicks on that button it shares their current location, picked from a list of physical addresses near the user, with the bot. This is useful for location-based services such as ride services or bots that facilitate pickups or deliveries.

Another type of canned response is provided by the Kik platform. Kik focuses on teens and simplifies the usage of buttons, restricting them to conversational navigation. Kik buttons are a type of canned response (called Suggested Responses) that the user can access from the bots keyboard, by tapping or clicking on the icon to the right of "Type a message" in the input field. The user can choose to use the given responses provided by the bot for this interaction, or revert back to the default keyboard to send free text to the bot.

In the example in Figure 9-19, the user is given a sweepstakes invitation and two options to reply. They can also click on the keyboard icon to the right of "Tap a message" to go back to a full QWERTY keyboard and send the bot free text.

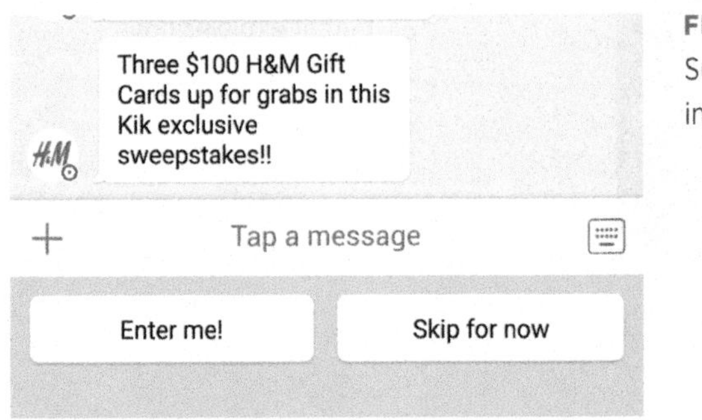

FIGURE 9-19. Suggested Responses in Kik

PUTTING IT ALL TOGETHER

A good example of combining image inputs, image outputs, and buttons is the Icon8 bot. Icon8 is a simple, task-led bot that helps you turn images into artwork.

At first the bot does a super-simple onboarding by showing what it can do—the bot has access to the user's profile photo and it takes it and automagically turns it into artwork (Figure 9-20).

FIGURE 9-20. Icon8 using images

The bot then goes on to post an animated GIF showing how a user uploads an image and gets an image back (Figure 9-21).

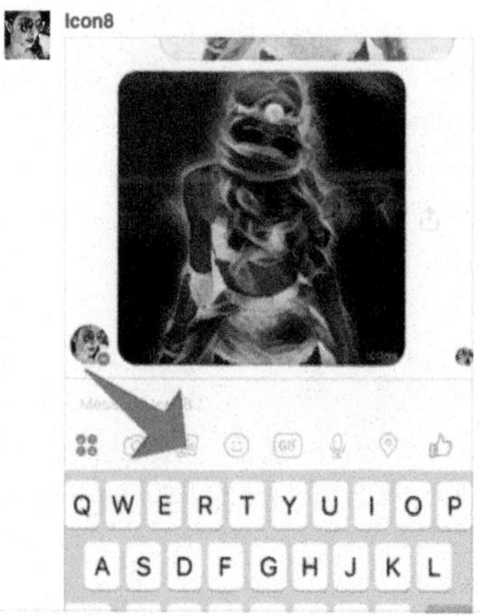

FIGURE 9-21.
The Icon8 bot posting a GIF that explains how to use the bot

This is a great "follow these steps" pattern that a lot of bot builders can use—seeing how the bot works in action is useful for understanding its value and how to work with it.

Lastly, the bot uses Quick Reply buttons to help the user navigate through the different filters the bot can overlay over an image (Figure 9-22).

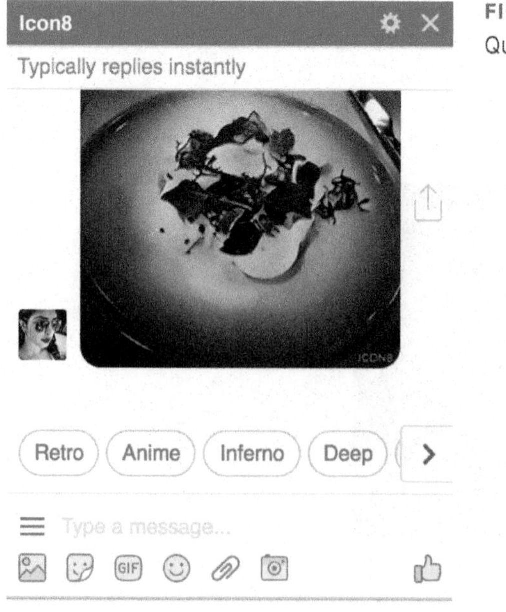

FIGURE 9-22.
Quick Replies in Icon8

Clicking on one of the buttons initiates the workflow, and the bot posts a beautiful version of your image. Unfortunately, the bot fails when the user provides an invalid input or asks for help (Figure 9-23). I guess it still has room for improvement.

FIGURE 9-23.
The bot is missing a help feature

WHEN TO USE BUTTONS IN A CONVERSATION

Buttons are a great navigational tool. In the last chapter we discussed the problems associated with the user getting lost in the conversation, the bot not understanding the user's intent, and the user diverging from the conversational task. Lead designers and product managers at both Facebook and Slack recommend using buttons to enable better conversation flow. Most of the new bots have moved to this navigation paradigm—but it is important to note that the user can still post free text to the bot, and that the bot should still apply logic to understand the user's input and to navigate to the right step in the conversation.

Buttons in Kik replace the keyboard, limiting the users' options and leading them to the viable responses. Buttons in Slack can also be used to facilitate workflows, provide canned responses, and act as in-app controls, where the user does not navigate through a conversation but rather through an app-like interface. Facebook Messenger provides a rich combination of buttons and canned reply controls that enable users to take action and also navigate the conversation with ease.

A great example of how buttons improved the design of a bot is told by Dan Manian, the cofounder of Donut (Donut is a bot that helps companies onboard, engage, and retain their employees):

Donut's first foray into the bot world was a simple Slack bot to help people to get to know their coworkers and teammates better. Donut regularly pairs up members of a Slack team who don't know each other well and invites them to get coffee, lunch, or donuts together.

We found out that if you ask users a yes-or-no question you can't necessarily expect them to answer with a simple "yes" or "no." Because it's a bot, if you ask something like, "Did you meet?" then people expect it to understand their response like a human would. We learned that if you want a structured response, like yes/no, then a more structured method of collecting the answers, like buttons, will work better and yield better data. So for example, before we had buttons we would ask a yes-or-no question like above, and if users said anything other than "yes" or "no" then we would respond and encourage them to say "yes" or "no." Since we were doing this in a group setting sometimes the users were having a conversation and after every message our bot would jump in and say "I didn't understand; please say 'yes' or 'no,'" which was a terrible experience:

@donut: Happy Friday! Did you meet this week?

@user_A: Wow, that's persistent!

@donut: Sorry, I'm not sure what that means. Please respond with something like `yes` or `no` or `yeah! totally Donut! you're the best!`

@user_A: 'Yes' ish

@donut: Sorry, I'm not sure what that means. Please respond with something like `yes` or `no` or `yeah! totally Donut! you're the best!`

@user_A: Yes

@donut: Great!

Switching to buttons solved this problem, and it increased our response rate by 20% because a button is just so easy to click [Figure 9-24].

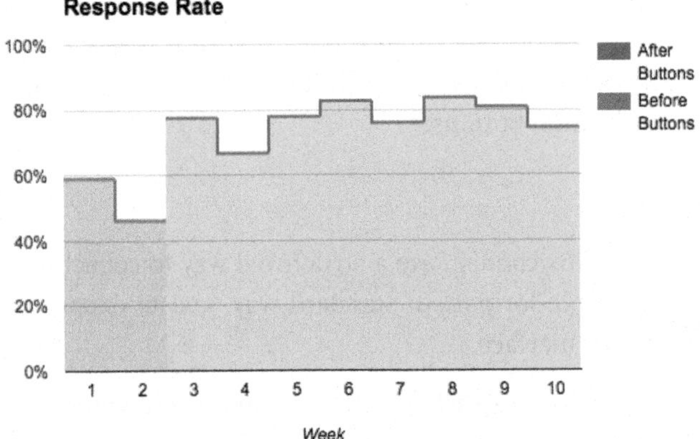

FIGURE 9-24.
Donut response rate before and after adding buttons

> For us buttons yield a much higher response rate than natural language question and answer, specifically for yes/no questions.

By switching from plain text to buttons, the team at Donut were able to work around the hard problem of understanding users' unstructured responses and provide a more productive user experience.

WHEN NOT TO USE BUTTONS FOR NAVIGATION

Buttons are not a magical solution that solves all of a bot's user input challenges.

Buttons are not a great user experience when there are a large number of options. Most platforms limit the amount of buttons you can add to a conversation in each interaction, and in any case you would not want your user to have to sift through a large set of buttons. Buttons are also not a valid choice when it comes to picking from an unknown set of options—emails, addresses, client names, and more. In these cases you need to default to free text and revert to the more complex intent detection and entity extraction.

Another example where buttons would not work is in free-form inputs, from describing how you are feeling to a coach bot, to providing expense justification to a finance bot. Anything where there is not a small, limited, and known number of options will not be a good fit for navigation through buttons.

Templates

Templates, in this context, are a structured way to collect different UI elements in a pre-formatted, standard way, and to expose these in a conversational interface.

Let's start with an example. Going back to the Kip shopping bot, Kip uses Messenger's *Generic Template* to display a carousel of items to the user in response to a search query (Figure 9-25).

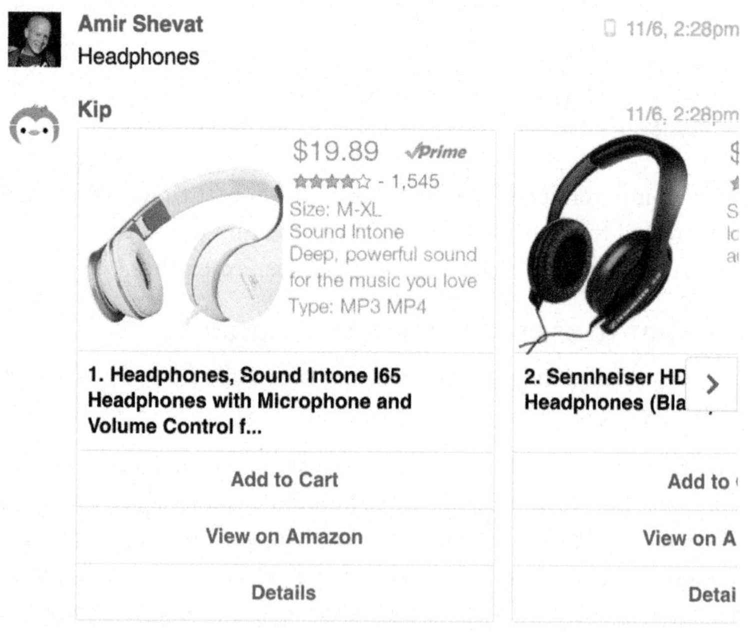

FIGURE 9-25.
In Messenger, Kip presents search results in a carousel

Note that the template has the image (the headphones with the price, rating, and a feature description in gray text), followed by the product description in black text, and a set of buttons under that.

In contrast, Slack uses a generic template style called *message attachments* where users can specify different attributes of the template (Figure 9-26).

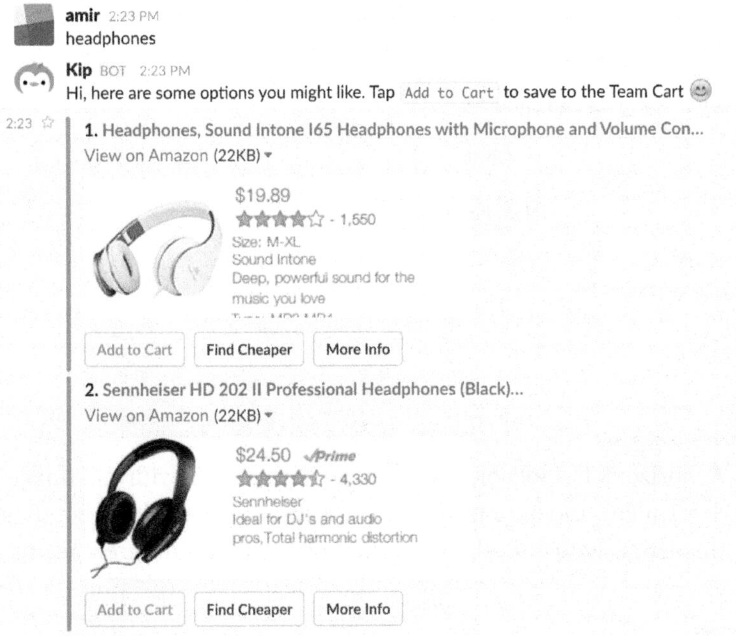

FIGURE 9-26.
In Slack, Kip uses a different template view to present the same results

As you can see, Kip provides the same information but in different template view, specific to Slack.

Now, let's dive a little deeper into templates.

Facebook Messenger provides templates that start with very generic and move fast to very specific use cases:

Button template
: A super-simple template that provides text and buttons. Figure 9-27 shows the Call of Duty game bot calling on the user to take the next step using this template.

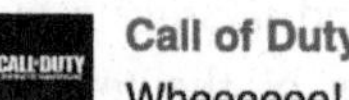

Call of Duty
10/28, 10:54pm
Hurry up and choose!
Nah, I'm good
Wheeeeee! You're up high! So before we all enjoy what you had for lunch, I'm going to turn the gravity back on.
Oh, silly me, one last thing! Remember to hold onto something. Gravity resuming in 3...2...

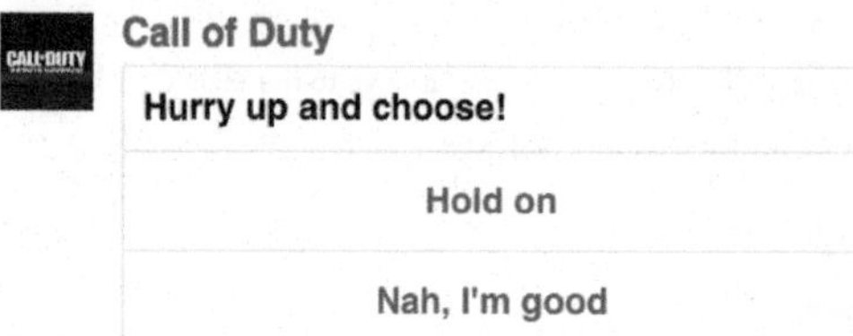

FIGURE 9-27.
The Call of Duty game bot implemented with the Button template in Messenger

Generic template

A horizontal scrollable carousel of items, each with an image, description, and buttons. Figure 9-28 shows how theScore bot uses this template to offer the user the choice to follow different teams.

11/11, 12:57pm
New on theScore: you can now follow College Basketball teams!

11/11, 12:57pm

FIGURE 9-28.
theScore bot implemented with the Generic template in Messenger

List template

A more condensed, vertical list of items, with optionally a cover image (Figure 9-29).

FIGURE 9-29.
An example of the List template in Messenger

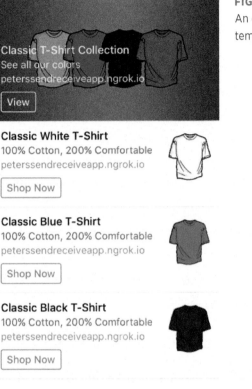

Receipt template

A dedicated template for bots providing commerce experiences. This is an easy way to provide a transaction summary and details.

Messenger also offers a set of airline-specific templates that define structure for boarding passes, itineraries, and so forth.

Developers and designers are free to use the templates for their own use cases. I have not seen many bots that use the more specific templates, but a lot of bots on Facebook Messenger take advantage of the more basic templates, such as the Button template for navigation and the Generic template for showing multiple items in a scrollable view. I am confident that more rich templates will be added in the future.

Slack takes a slightly different approach to templates—the Slack platform provides bot builders with a single, customizable template called the message attachment. Message attachments are highly configurable and can cover a large set of use cases. Figure 9-30 shows just one example of this template.

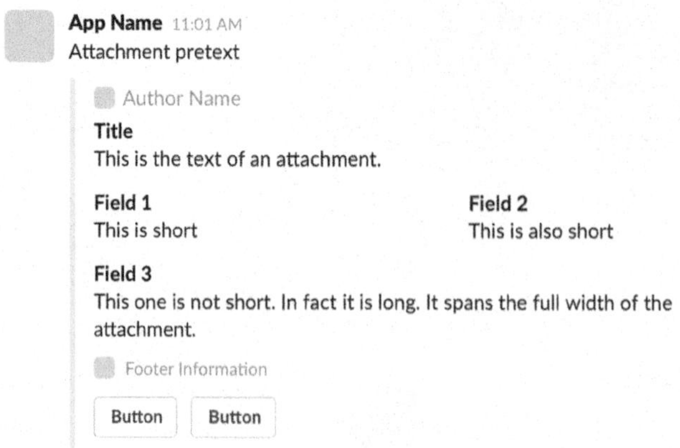

FIGURE 9-30.
Basic message attachment template in Slack

The message attachment is a generic template that provides a consistent look and feel for Slack users. Note that Slack renders the attachment differently on mobile and desktop, so make sure to test on different platforms.

In each interaction, the bot can send multiple attachments to the user, each with a different set of information and buttons. Bot designers can customize each field and provide text formatting and color-coding in each attachment. In Figure 9-31 you can see how the Beep Boop DevOps bot indicates that one of my bots is offline with a red color.

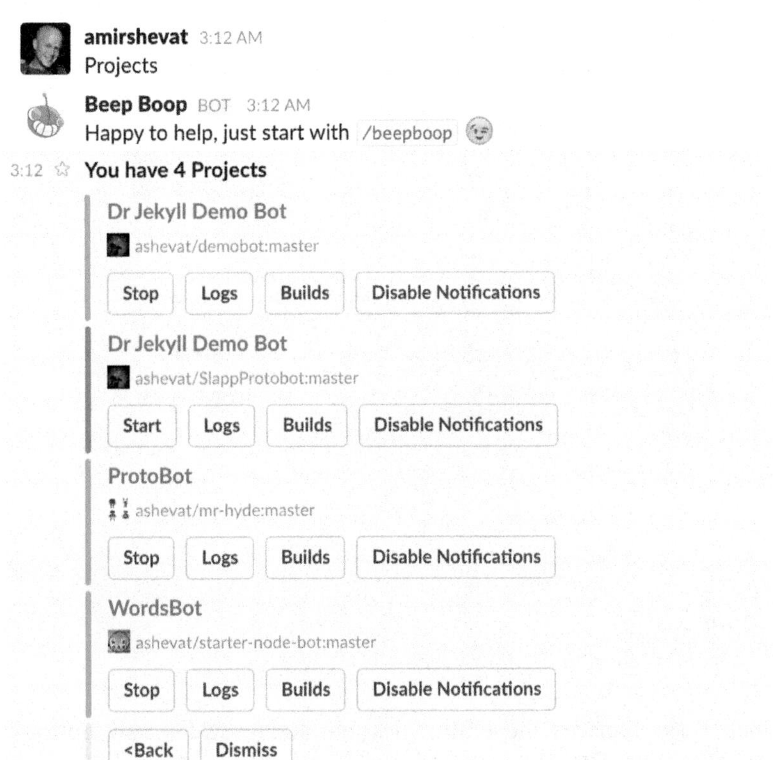

FIGURE 9-31.

An app-like interface using templates and buttons

Remember that when a user clicks a button, the bot can alter the content of the message attachment, creating an app-like experience. Clicking on the "Start" button for the offline bot will change the color to green, and clicking on "Logs" will replace the entire message attachment with the log view (Figure 9-32).

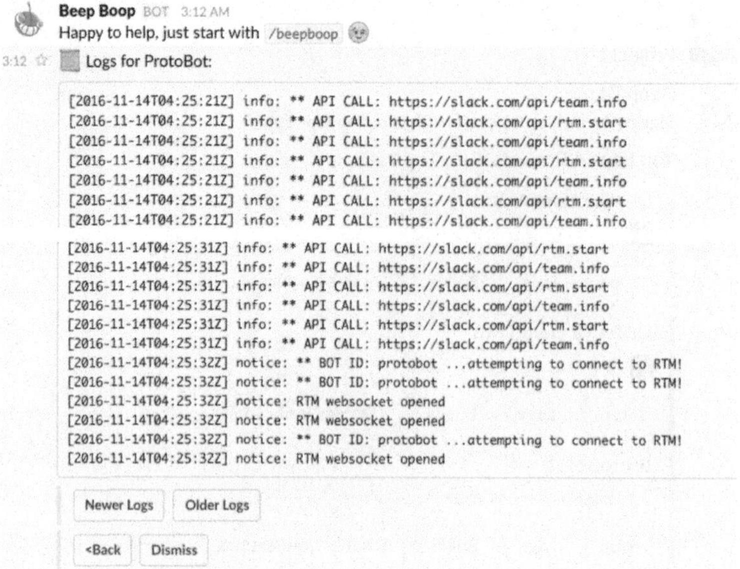

Beep Boop BOT 3:12 AM
Happy to help, just start with /beepboop
3:12
Logs for ProtoBot:
[2016-11-14T04:25:21Z] info: ** API CALL: https://slack.com/api/team.info
[2016-11-14T04:25:21Z] info: ** API CALL: https://slack.com/api/rtm.start
[2016-11-14T04:25:21Z] info: ** API CALL: https://slack.com/api/team.info
[2016-11-14T04:25:21Z] info: ** API CALL: https://slack.com/api/rtm.start
[2016-11-14T04:25:21Z] info: ** API CALL: https://slack.com/api/team.info
[2016-11-14T04:25:21Z] info: ** API CALL: https://slack.com/api/rtm.start
[2016-11-14T04:25:21Z] info: ** API CALL: https://slack.com/api/team.info
[2016-11-14T04:25:31Z] info: ** API CALL: https://slack.com/api/rtm.start
[2016-11-14T04:25:31Z] info: ** API CALL: https://slack.com/api/team.info
[2016-11-14T04:25:31Z] info: ** API CALL: https://slack.com/api/rtm.start
[2016-11-14T04:25:31Z] info: ** API CALL: https://slack.com/api/team.info
[2016-11-14T04:25:31Z] info: ** API CALL: https://slack.com/api/rtm.start
[2016-11-14T04:25:31Z] info: ** API CALL: https://slack.com/api/team.info
[2016-11-14T04:25:32Z] notice: ** BOT ID: protobot ...attempting to connect to RTM!
[2016-11-14T04:25:32Z] notice: ** BOT ID: protobot ...attempting to connect to RTM!
[2016-11-14T04:25:32Z] notice: RTM websocket opened
[2016-11-14T04:25:32Z] notice: RTM websocket opened
[2016-11-14T04:25:32Z] notice: ** BOT ID: protobot ...attempting to connect to RTM!
[2016-11-14T04:25:32Z] notice: RTM websocket opened
Newer Logs
Older Logs
<Back
Dismiss

FIGURE 9-32.
Beep Boop displaying the log view for one of my bots

Beep Boop replaces the entire message and provides new buttons, offering additional navigational and functional controls.

In a different use case, Figure 9-33 shows how Statsbot, a marketing analytics bots for Slack, uses message attachments to display a report with a chart (in an image format).

amirshevat 8:29 PM
users yesterday
statsbot BOT 8:29 PM
Google Analytics users for Nov 12
Click here for full report (beta)
22 users
↓ 31.25% (32)
(46KB)
Schedule it

FIGURE 9-33.
Statsbot displaying a chart in a message attachment

Notice that the button at the bottom of the message attachment has a call to action to schedule this report—this is a great way to ensure user reengagement. Also note the link at the top of the report, sending the user to the full detailed report on the web.

In Slack, the only way to display buttons as part of a conversation is with a message attachment. A message attachment can be composed solely of buttons, and a message attachment with buttons can serve as a navigational aid. Figure 9-34 shows how Sensay uses this feature.

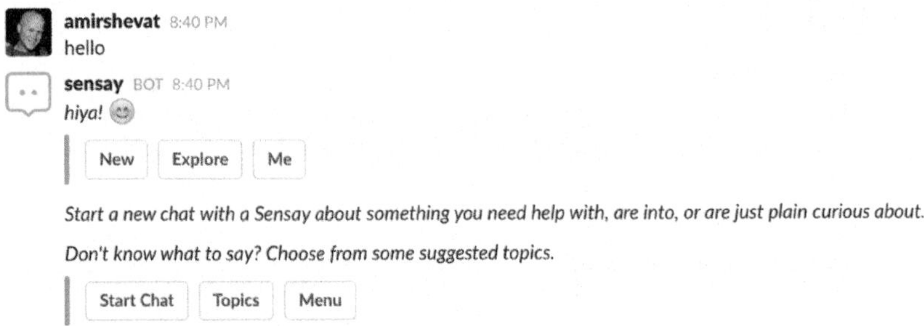
amirshevat 8:40 PM
hello
sensay BOT 8:40 PM
hiya!
New
Explore
Me
Start a new chat with a Sensay about something you need help with, are into, or are just plain curious about.
Don't know what to say? Choose from some suggested topics.
Start Chat
Topics
Menu

FIGURE 9-34.
Sensay using attachments and buttons as navigational aids

WHEN TO USE TEMPLATES IN A CONVERSATION

As you can see, templates are a good way to organize complex information and a rich set of controls in an app-like context. Users can interact with the templates in a familiar way, because they are consistent across all bots on the platform. Generic templates in Messenger and Slack provide a good way to display lists of items and their associated actions. Each item in the list is an object in the user's mental model, and the buttons represent tasks or actions that can be performed on the object.

Links

Most platforms support links. Links are an easy way to send the user out of the conversation and into the web. Links also serve as a way to refer to something on the web and surface a preview of it in the conversational interface—technically this is referred to as *unfurling*.

Sending a user to a web page can serve a few use cases:

Performing out-of-conversation actions, like authentication with third-party services
: Sometimes your bot needs to run an authentication flow that is only available on the web (using OAuth, for example). In this case (Figure 9-35) the bot sends a link to the user and asks them to authenticate.

amirshevat 5:59 PM
report
Help-desk BOT 5:59 PM
We need to authenticate you within the corporate helpdesk center.
Please use the following link to authenticate - http://www.foo.bar/auth

FIGURE 9-35.
Using links for authentication

This is a recommended pattern in Slack—providing a username and password for a third-party service inside the chat client (for authentication with other systems, for example) is highly discouraged.

Expanding on the given information
: Bots can provide a summary of a report (or a search result) and send the user to the browser for the extended report. Figure 9-36 shows how Statsbot does it.

FIGURE 9-36.
Using links to direct users to expanded information

Statsbot pulls out the summary and key performance indicators, and posts them in the conversation—but when it comes to drilling down to see the full report, the bot posts a link to Google Analytics. This pattern of providing a summary of content in the chat interface is useful because space is limited, and in most cases users do not need more than the summary. Statsbot actually solves a lot of the user's needs by providing a short summary, without the complexity of a lengthy report.

Promoting web content

Some bots provide limited content and drive most of the engagement into the web. The key is to provide enough value in the chat interface so the user does not get the feeling of a shallow bot. As with mobile apps that serve as gateways to web pages, there is a risk that bots that provide low value in the conversation will be considered of low quality.

Unfurling

Another major use case is around links that facilitate the "importing" of content from the web into the conversation. Unfurling is the technical term for what happens when you share a link on Twitter, Facebook, LinkedIn, and Slack. You might notice that these services provide a preview/summary of the page inline in your posts. This is useful because the summary provides enough information to let the user understand what the link is all about, so they can decide whether they want to take a closer look.

In Figure 9-37, I posted a link to *api.slack.com* and Slack unfurled the link and provided a preview of that page. This is a standard process where the platform pulls out selected HTML tags embedded in the web page and formats them into the conversation.

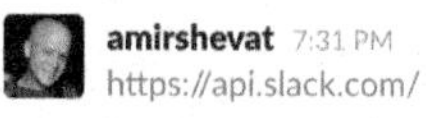

Slack API
Slack APIs allow you to integrate complex services with Slack to go beyond the integrations we provide out of the box.

FIGURE 9-37.
Unfurling in Slack

In the same way humans can post links that unfurl, bots can do that too in Slack (Figure 9-38).

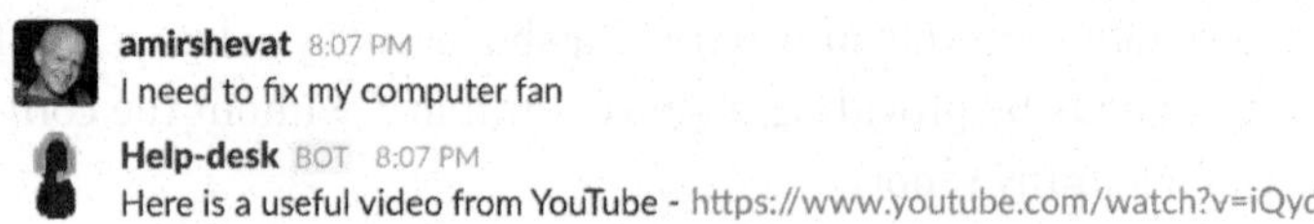
amirshevat 8:07 PM
I need to fix my computer fan
Help-desk BOT 8:07 PM
Here is a useful video from YouTube - https://www.youtube.com/watch?v=iQydtO0l0as
YouTube | jaykay18
Computer Fan Repair--THE EASY WAY!!!

FIGURE 9-38.
Bots can post links that unfurl too

As you can see, the bot did not upload a video, it just posted a link to YouTube—and the video, together with its player, appeared inline in the conversation.

Another interesting use case with unfurling is the *authenticated unfurl*. Think about posting a link to a website that requires authentication, such as a link to an account in your CRM. In this use case, the bot might require a special process, using the posting user's authentication token, to unfurl the URL. This is especially common when connecting to a secure/internal line-of-business web application.

WHEN TO USE LINKS IN A CONVERSATION

Use links to send the user to the web from the conversational interface—this could be to complete a workflow that cannot or should not be accomplished in the conversational interface, or to direct the user to content better consumed on the web. In some platforms, you can also use links to enrich the conversation with content from the web that is unfurled into the conversation itself.

Emojis

Emojis and emoji reactions are becoming a common way to rely emotions and information, and even to take action and denote a process. Figure 9-39 shows a common pattern in internal communication at Slack.

FIGURE 9-39.
Using emoji reactions to convey responses

These four simple emoji reactions relay a lot of information: I agree, I am looking at it, it's done, and I'm very happy it is resolved!

Emojis are also useful for emphasis—you can add an emoji of a fire next to a burning issue to stress the fact that it is urgent or important.

Emojis can be inlined into an everyday conversation too (Figure 9-40).

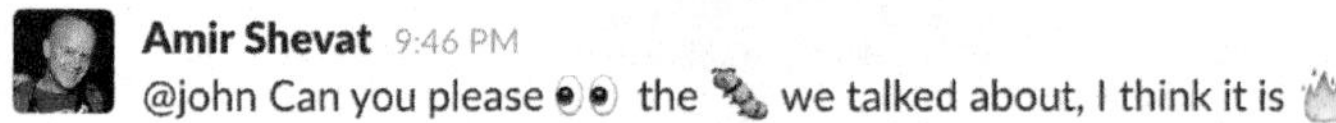

FIGURE 9-40.
Using emojis in conversation

While this specific example is an extreme case of emoji usage, and probably not the recommended way to pass this information, most bot developers report seeing emojis in the conversational input from users. This is now a standard part of most mobile keyboards and is even supported in SMS and email clients.

Bots can also post emojis in most chat platforms. The ability to add emoji reactions is currently limited to Slack, both on the bot side and the user side.

It is pretty delighting to get a ☑ from a bot when you ask it to perform a task. Bots can also register to be notified when users add an emoji reaction to a message, and implement processes utilizing these reactions.

Figure 9-41 shows an example of how a bot can use emojis to entice the user to vote on a topic.

Help-desk BOT 10:00 PM
Lets go out for lunch! Please vote or .

FIGURE 9-41.
Voting with emojis and priming the user to pick a pizza or a burrito

Users in the channel can each add an emoji reaction to that post and the bot can decide, say after 30 minutes, to close the count and announce the destination for lunch. Note that this practice was somewhat deprecated when buttons were introduced into the platform, but it is still used by some bots.

Another interesting example of extreme use of emojis is the Emoji News bot, a bot that sends you news composed mainly of emojis (Figure 9-42).

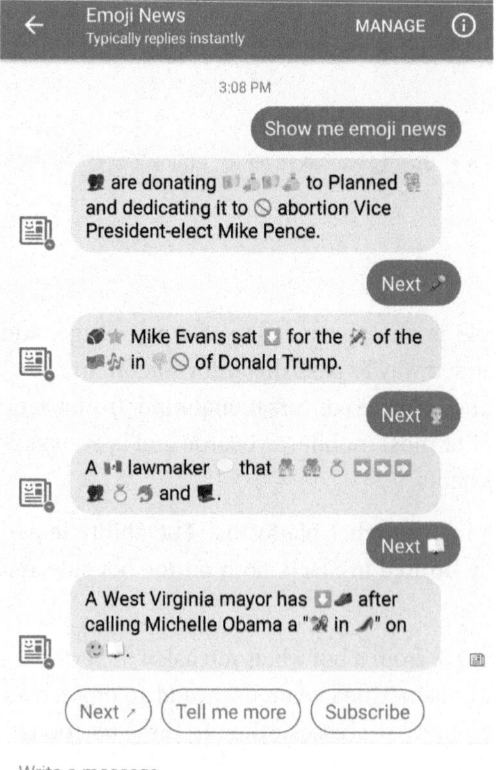

FIGURE 9-42.
The Emoji News bot

The attraction here is the riddle-like interface: you need to decipher the intent of the emojis in the context of the news headline. The users play a game with themselves to see if they can understand each title, in which many of the words are replaced with emojis.

WHEN TO USE EMOJIS IN A CONVERSATION

Use emojis to relay information, enrich the conversation, and indicate actions taken. Since the introduction of buttons it is not recommended that you implement a full process with emojis only, but implementing a few delightful interactions using emojis might be a great alternative to using text exclusively.

Typing Events

In several platforms you can send a "typing" event on the bot's behalf, which is usually visualized inline in the chat app. This is useful to give the user the impression that the bot is working on a given task or typing a long post. Users are sometimes annoyed when they do not hear back from a bot quickly, as they feel the bot is not responsive. Sending the "typing" event can give the user the comfort that the bot is working on a reply.

Conversely, some users get unsettled when the bot answers too fast. Coming back to the UX mental model of chat, it feels unnatural to some users that the bot is not taking its time to type the message. Using typing event together with a 1–2 second delay is a way to give users the feeling that they are conversing in a more natural environment.

Persistent Menus

The Persistent Menu control is currently unique to the Facebook Messenger platform. It lets you add a menu that is persistent throughout the conversation with your bot. The concept is interesting because it can potentially be used as a top navigation control. A user can access the menu at any time via the burger-like icon next to the text input (see Figure 9-43).

FIGURE 9-43.
The persistent menu icon (left)

Clicking or tapping on the burger icon pops up a menu with the items set by your bot (Figure 9-44).

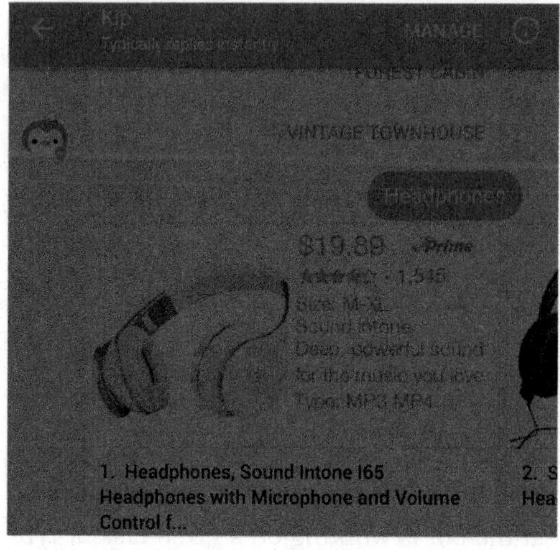

FIGURE 9-44.
Displaying the menu

Currently bot designers are reporting low user engagement with this persistent menu, perhaps because users are not yet familiar with the functionality. So, the current recommendation is to put lower-priority action items in this menu, such as Help and Feedback.

At the time this book is going to print Facebook Messenger is exploring a new user experience for its menus (currently only available in the mobile version of Messenger), and early reports are showing good user engagement. As part of this new user interface developers can limit the user's ability to type free text, and confine the user to picking from Quick Replies and buttons only.

Slash Commands

Similar to menus, slash commands are a way for a bot to add functionality inline in the chat app itself. Slash commands are command line–like actions that autocomplete when a user types a "/" in the chat app input box (Figure 9-45). They are currently only supported on the Slack platform.

FIGURE 9-45.
Slash command autocomplete list

Bot designers can implement their own slash commands that are added to this autocomplete list. As you can see, you can also set short descriptions and usage hints that show in the autocomplete list.

Typing the slash command in Figure 9-46 will result in a response from the Appear.in bot (Figure 9-47).

/appear bots-are-cool

FIGURE 9-46.
Typing this slash command in Slack...

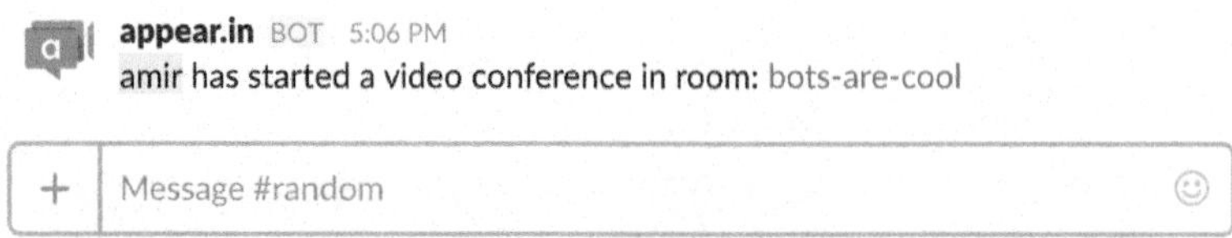

amir has started a video conference in room: bots-are-cool
Message #random

FIGURE 9-47.
... produces this result

Appear.in is a simple bot that creates a virtual meeting link, which you can specify with this simple Slack command.

Slash commands can also be a good way to show information, in a team context, to only the user who ran the command. Slack calls this an *ephemeral message*. Bot designers can specify whether the response to a slash command will be visible to everyone or just the user who invoked the slash command.

Figure 9-48 shows an example of an ephemeral message.

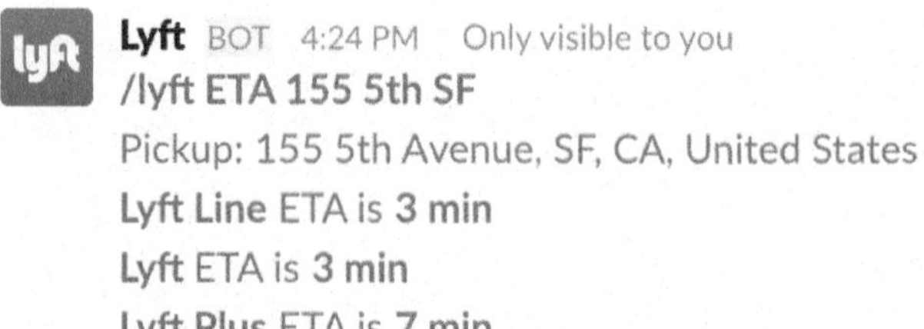

lyft
Lyft BOT 4:24 PM Only visible to you
/lyft ETA 155 5th SF
Pickup: 155 5th Avenue, SF, CA, United States
Lyft Line ETA is 3 min
Lyft ETA is 3 min
Lyft Plus ETA is 7 min

FIGURE 9-48.
An ephemeral message in Slack

Note the wording at the top that says "Only visible to you." Using ephemeral messages is a good pattern when you want to avoid starting a direct private conversation with a user, but still provide that user with information only they can see.

Like menus, slash commands are not used heavily by most users. A command-line paradigm lends itself more to a technical audience. Slash commands are good for enabling a shorthand version of a conversation, however. In the example in Figure 9-48, by inputting "/lyft ETA 155 5th SF" the user provides both the intent (to get an estimated time of arrival) and the entity (the address, 155 5th Street, San Francisco) that is required to accomplish the intent.

Webviews

Webviews are provided by the Kik and Facebook Messenger platforms as a way to expose a web browser and direct the user to or augment the conversation with web content. For example, the user can play a conversational game on Kik and then be routed to a more interactive webview action part of the game (where you crush zombies with your fingers, for example). In Messenger, the use cases are typically around extending and enriching an ecommerce conversation.

In many cases a webview is just a link to a page on the service's website. In use cases like that, promoted by the Facebook Messenger platform, the bot serves as a gateway or additional pointer to the website. Figure 9-49 shows an example.

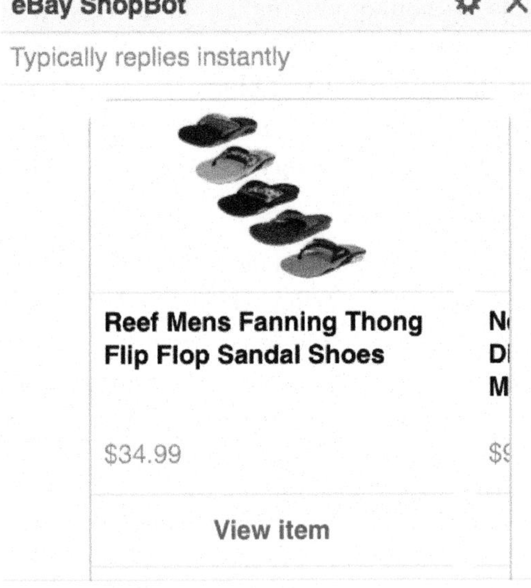

FIGURE 9-49. A webview on Messenger

In this example, the eBay bot provides the capability to search the eBay inventory, but at the end of the process the user is sent to the website to view the details and complete the transaction.

Connecting It All Together

Rich elements provide you with a way to augment the conversation with visual information, provide users with controls that help them navigate through the conversation, and structure information in a well-known and templated way. Not every interaction requires a conversation. Some tasks can be fulfilled with simple rich interactions.

[KEY TAKEAWAY]

Not every interaction requires a conversation. Some tasks can be fulfilled with simple rich interactions.

The major risk with overusing rich controls is that as designers we will default to the old mobile/web ways and do very simple ports of our mobile or web services into the chat interface. This is a very common mistake—porting a service from one platform to another (from Android to iOS, for example) or from one UX paradigm to another (for example, from web to mobile) without modifying it to fit the new environment always yields an inferior user experience. If you are just doing a dumb port of your website to a bot, users will prefer the web experience, and if you are copying all the buttons in a mobile app into a bot conversation, you are probably providing a very cumbersome experience for your users.

From a navigational point of view, buttons are a useful component that can guide users though the conversation, prime them to pick the right choice (with color-coding and emphasis), and direct them back to the happy path when they are lost. Bots cannot block users from entering free text, though, so remember the users can always opt to ignore your buttons and enter their own input.

The key is to understand how your users expect to interact in a conversation, provide just the right information at the right time, and use rich elements to facilitate the interaction when applicable.

The best bots will be a hybrid of text and rich interaction; they will provide the user with a delightful conversation enriched by rich controls. Great bots will provide buttons but support text responses, will post summaries and provide links to details, will expect rich interaction and feedback from the user's end, and will add a layer of sentiment with emojis, images, and GIFs.

Here is an interesting insight shared by Dennis Yang, cofounder of Dashbot:

> Affordance is a known design principle that definitely still applies to chatbot conversations, and we would be remiss to overlook it. Most of these lessons that we've learned, thus far, are around understanding the unique affordances that chat provides for your users. Buttons and menus are great ways to explicitly define what is possible for your user to do, and the text box offers nearly an infinite spectrum of options for your users. So, tailoring your bot experience to optimize and handle the options afforded by chat is a great way to approach bot design.

Closing Thoughts

This chapter has explored the rich controls that are currently available in the different bot platforms, from images and videos to buttons and Quick Replies. In the near future we will likely see the various platforms releasing more and more rich controls—this is their way to provide designers with an alternative to plain text inputs, as well as to enrich the experience on a particular platform. As designers it is our job to take the building blocks provided by the platforms and try to create the best-in-class experience for our users.

In the next chapter we will explore an important aspect of our bots' interactions—we will take a deep dive into memory and conversational context, which are the pillars of every intelligent human conversation.

[10]

Context and Memory

Memory... is the diary that we all carry about with us.

—OSCAR WILDE

MEMORY AND CONTEXT ARE a natural part of every human conversation. We have an implied understanding that our counterpart in the conversation remembers what we said a minute ago, or that they know that we are now talking about cars, or that they have an understanding of the thread of thoughts tied into a conversation. All of this is trivial for the human mind, but extremely hard for software.

Bot Amnesia

Here is an example of how bots fail in this simple task. Figure 10-1 is a humoristic gift that was given to every participant in a bot event in Europe this year. It symbolized the main problem that bot conversation faces today.

FIGURE 10-1. Bots often have trouble maintaining context

Many bots today focus on a request/response paradigm. In this paradigm each request has a new context, and all past contexts are forgotten. Some bots do a better job of maintaining context than others—for example, the Google Assistant bot isn't bad (Figure 10-2).

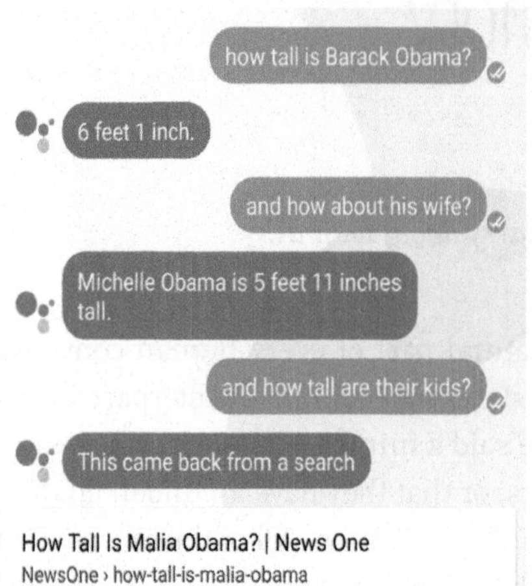

FIGURE 10-2. Google Assistant does a decent job of following the thread of a conversation...

As you can see, the user in this example uses words like "his wife" and "their kids." For us humans, it is trivial to understand that "his wife" refers to Barack Obama's wife and that "their kids" refers to the Obamas' kids. But disambiguating that in a conversation with a bot is not that easy, and even Google Assistant fails in some use cases (Figure 10-3).

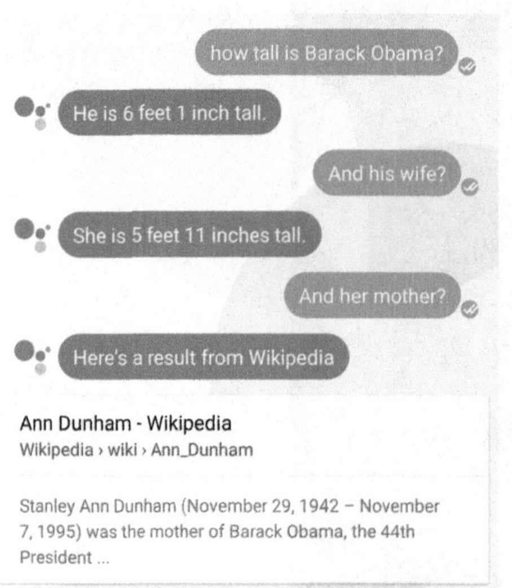

FIGURE 10-3. But even it fails in some cases

The context here started with Barack Obama, but then it moved to his wife. To a human, it's clear that "her mother" is referring to Michelle Obama's mother. Notice that there is no ambiguity because of the gender associated with the question; "her" plainly refers to Michelle and not Barack. But Google Assistant kept the first context and returned results associated with Barack's mother.

Other forms of bot amnesia include forgetting the user's name, address, preferences, and more. The human brain can hold a full conversation in memory. We can be talking about a trip to Rio, divert to a conversation about Indian food, then revert to the trip to Rio, with all the associated context, with ease. Next we will explore the essence of context and memory, and we'll see how we can cure our bots from this amnesia and even create delightful moments of recollection.

Context

One way to look at context is as the intent and the set of variable entities associated with the current conversation. When requesting a vacation from an HR bot, the context might look like this:

Intent: Paid time off

Entities:

User: Jassim Latif

Start Date: 04/07/2017

End Date: 04/09/2017

Some of these context variables are *scoped*, or local, to the intent (such as start date and end date), and some of the context variables are *global* (like the user). Scoped variables are more tightly coupled with the intent, so when a user moves to another intent, such as commuter benefits, these variables can be forgotten. Global context variables are variables that the bot should remember across all intents.

In some cases, scoped context variables can be useful across related intents—for example, if a user is going on a business trip and initiates a conversation with the intent *book flights*, it would be super nice if, when the user's intent changes to *book hotels*, the bot could remember the dates scoped to the last intent and offer them as options for that intent:

User: Book a flight

Travel-bot: Where would you like to go?

User: Rio Brazil

Travel-bot: On which dates?

User: 4-6 of November 2017

...

User: Book hotel

Travel-bot: Would you like me to book a hotel in Rio Brazil, 4-6 of November 2017?

User: Yes! Thanks!

Global variables should always be consistent across intents, and the bot should remember these and offer them as defaults. Another way to think of this is global variables as the long-term memory and scoped variables as the short-term memory—but while humans scope their memory with time, this is not applicable to bots, which can remember things indefinitely.

When possible, bots should not time out context—if a user starts an intent, they should be able to come back to it after a while and the bot should persist (remember) the context. It is really aggravating to start a process, get interrupted, and then have to provide all the information again just because a few hours have passed. This is especially true in text-based conversations where immediate replies are not implied. In text-based conversation it is common to continue a conversation when you are available or the next time you are online.

[KEY TAKEAWAY]

When possible, bots should not time out context—if a user starts an intent, they should be able to come back to it after a while and the bot should remember the context.

There are rare occasions when global variables should be forgotten, for compliance and legal reasons, but in most cases there is an implied understanding that the bot will remember the users and their details.

For complex use cases, an intent itself can be part of the context of another intent. Consider the example in Figure 10-4.

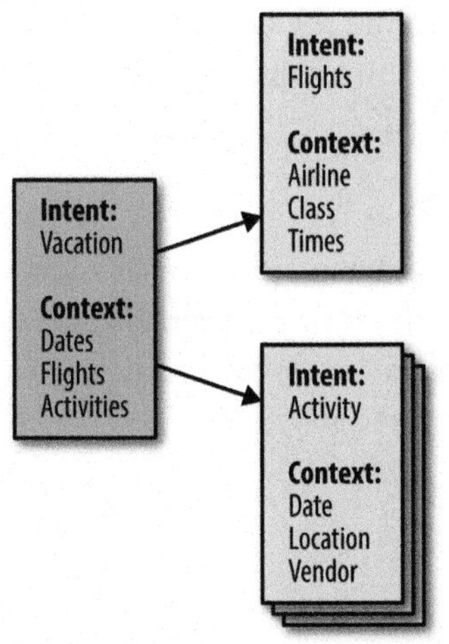

FIGURE 10-4.

An intent can form part of the context of another intent

As you can see, each intent has its own context, but some intents are nested in the context of another intent. The hierarchy of intents can be much more complex than this example, with multiple levels of nesting—for instance, picking the seats for a flight can be a subintent of the flight intent.

The bot should navigate the user through the path of the intents, but should also support the user with the ability to traverse up and down the intent tree:

> ...
>
> ***Travel-bot***: Great, you are booked for bungee jumping on Oct 3rd at 4 p.m.
>
> *User*: Fantastic. Remind me of our return flight time? We might want to do something on the 4th.
>
> ***Travel-bot***: Your return flight (Trip to New Zealand) is at 6 p.m. on Oct 4th.

The bot was able to understand the context of the request: it traversed up the intent hierarchy and pulled a scoped context variable from the flight intent. The bot kept the master context of the vacation and implied the flight context from it.

If the user had instead said "What are our flight details for the Christmas vacation to Cancun?" the answer should have been very different. The strategy that is most effective in many cases like this is to try to traverse up the intent tree until you find the closest intent that matches the user's input. In some use cases this might not be so easy, and additional logic might need to be applied.

INFERRING CONTEXT FROM PRONOUNS

Another big challenge is inferring what pronouns in the user inputs are referring to. Users use pronouns all the time to refer to context variables. Words like "his," "hers," "this," and "it" serve as pointers in a conversation, pointing to particular context variables:

> ...
>
> *User*: When is my meeting with John Agan?
>
> ***Travel-bot***: Oct 3rd at 4 p.m.
>
> *User*: Can you please move **it** to 4 p.m.?

The word "it" in this example is pointing to the meeting. It is expected that the bot *remembers* that variable and infers it from the context.

This is part of the natural language understanding (NLU) problem domain—this process can be done by parsing the sentence using traditional algorithms, or using an artificial intelligence system with NLU capabilities. As we saw in Chapter 7, the ability to understand and infer context in complex, multi-intent conversations is sometimes called *conversation management*; while more complex, it uses the same basic concepts.

There are many AI frameworks that promise to solve this challenge through one means or another. From a design perspective, you do not really have to worry about which underlying technologies are used, as long as they provide you with the right outcomes. As a designer, you will need to provide the conversational flow and several permutations of each user input. An example design is discussed in Chapter 16.

DERIVING CONTEXT THROUGH RICH CONTROLS

Exposing the context (intent and entities) with rich controls is another way to clearly capture intent. Figure 10-5 shows how Google Assistant does this.

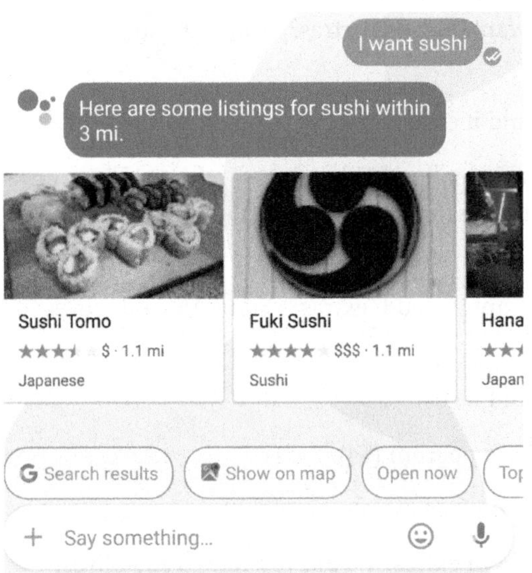

FIGURE 10-5. Google Assistant offers rich controls to help derive context

Google Assistant offers the user contextually relevant buttons that can drive the conversation forward. In this example, the user intent was to have sushi; Google served relevant search results, followed by an offer to show the results on a map, or filter the results to better fit the user's preferences.

Not all bots require context—some bots are designed for simple request/response interactions. For example, my WordsBot has a simple "define word" → "word definition" paradigm; the bot does not remember any past transactions and is not expected to.

> **[KEY TAKEAWAY]**
>
> Not all bots require context—some bots are designed for simple request/response interactions.

If you are interested in providing a bot that supports rich conversation, at the most basic level, the bot should remember a set of local and global variables and be able to infer intent and the referents of basic pronouns, providing the user with a natural and intuitive conversation.

CONTEXT ERRORS AND AMBIGUITY

So what happens if you get the context wrong? This can happen to the best of bots, and can be driven by human error or language ambiguity:

> ***Pizza-bot***: Do you want to add extras? You can add olives or pepperoni.
>
> *User*: How much would **it** cost?

What would be the right answer here? What does *it* refer to? There are a few strategies to disambiguate this conversation. One way is to provide an answer like this:

> ***Pizza-bot***: Do you want to add extras? You can add olives or pepperoni.
>
> *User*: How much would it cost?
>
> ***Pizza-bot***: Olives ($2), Pepperoni ($4)

You can also ask about each pizza extra separately:

> ***Pizza-bot***: Let's talk about extras. Would you like to add olives?
>
> *User*: No
>
> ***Pizza-bot***: Pepperoni?
>
> *User*: How much would it cost?
>
> ***Pizza-bot***: $4
>
> *User*: OK, let's do that.

There are a lot of other ways to resolve disambiguation, and they are all context-dependent. Figure 10-6 shows an example of a conversation I had with Mitsuku, a delightful conversational bot on the Kik platform. It got confused, admitted it, and moved on.

Because Mitsuku is not a mission-critical bot but a topic-led chitchat bot, it can more easily recover from contextual errors. Topic-led bots, in general, can recover from context loss by changing the conversation topic, or moving on with it.

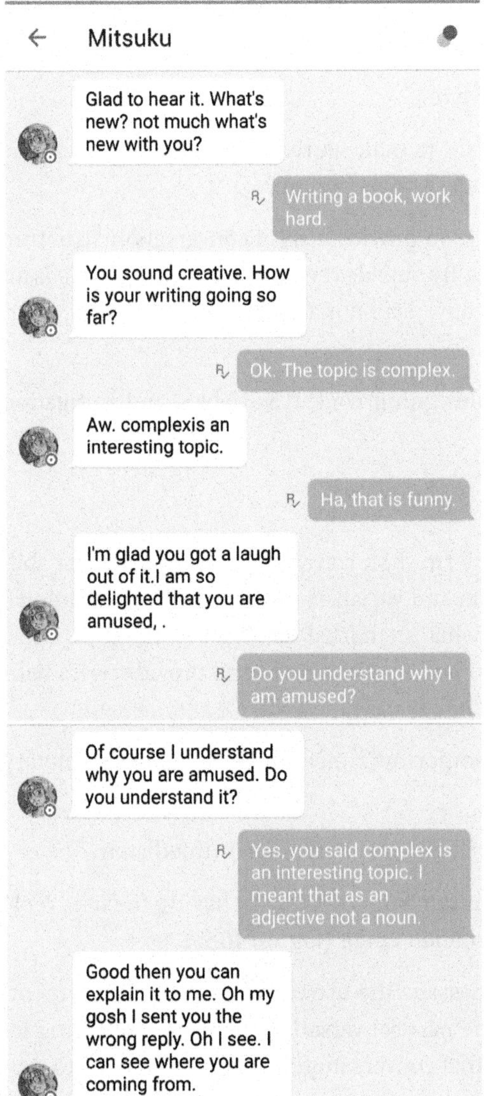

Mitsuku
Glad to hear it. What's new? not much what's new with you?
Writing a book, work hard.
You sound creative. How is your writing going so far?
Ok. The topic is complex.
Aw. complexis an interesting topic.
Ha, that is funny.
I'm glad you got a laugh out of it.I am so delighted that you are amused, .
Do you understand why I am amused?
Of course I understand why you are amused. Do you understand it?
Yes, you said complex is an interesting topic. I meant that as an adjective not a noun.
Good then you can explain it to me. Oh my gosh I sent you the wrong reply. Oh I see. I can see where you are coming from.

FIGURE 10-6. Mitsuku gracefully recovering from a context error

In task-led conversation, however, the bot needs to collect all the relevant context variables to complete the transaction. When a task-led conversational bot has a low degree of confidence that it has the right context, it should revert to error flows such as course correction or seeking human intervention (see Chapter 8). Sometimes just having the bot admit that it does not understand and doing a simple course correction (or requesting confirmation) might set the bot back in context:

> ***Meeting-bot***: You have meetings with Stewart Butterfield at 3 p.m. and April Underwood at 4 p.m.
>
> *User*: Let's move it to 5 p.m.
>
> ***Meeting-bot***: Sorry, I want to make sure. Do you want to move the meeting with April Underwood to 5 p.m.?

Another common strategy is to provide a rigid conversation structure and limit the user functionality outside the happy flow. While this is an easier strategy in terms of conversation management, it is not necessarily the most user-friendly.

We will talk more about designing context variables and mitigation flows in Chapter 16.

Memory

Memory is a more generic term than conversational context. The ability of bots to pull up intents and variables, or even full conversations, from the past is still somewhat uncharted territory. At the most basic level, bots should be able to pull up old intents and provide users with the ability to act on them:

> *User*: I need to move tomorrow's meeting with Taylor Singletary to Friday
>
> ***Meeting-bot***: Roger! I will send him an email immediately.

The scheduling bot Amy Ingram supports this by having the user reply to the original meeting invitation email (Figure 10-7).

By replying to the initial message the user can "refresh" Amy's memory and provide a link to the past conversation the user is referring to. Amy then has a pointer to that conversation, so there is no need to provide further context.

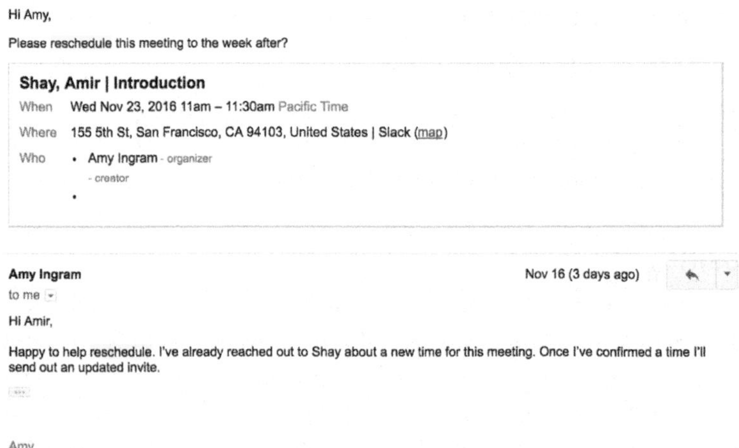
Hi Amy,
Please reschedule this meeting to the week after?
Shay, Amir | Introduction
When Wed Nov 23, 2016 11am – 11:30am Pacific Time
Where 155 5th St, San Francisco, CA 94103, United States | Slack (map)
Who • Amy Ingram - organizer
- creator
Amy Ingram
Nov 16 (3 days ago)
to me
Hi Amir,
Happy to help reschedule. I've already reached out to Shay about a new time for this meeting. Once I've confirmed a time I'll send out an updated invite.
Amy

FIGURE 10-7.
Amy uses the original email as the context of the conversation

Providing a memory of transactions and entities that can be accessed is useful not only for functionality's sake, but also for building trust between the user and the bot. Users need to feel that they can go back and edit/modify/cancel or just revisit past transactions. Supporting functionality like "list projects," "view past meetings," or "load last pizza orders" can help bestow trust in the user's mind. Remember that bot users do not have top navigation menus, so giving them a way to access history serves an important functional need.

Another aspect of memory is association. This is very common in topical conversation; we talk about something, and it reminds us of another topic. Bots can use this attribute to naturally navigate the conversation to another place:

> ***Starwars-bot***: Speaking of Yoda, this **reminds** me that you can now buy his new figure toy in our online store. Would you like to go there?

Here, the bot makes an artificial memory association that steers the conversation down the funnel of buying merchandise. Recollection of past conversations is also a good way to reengage with the user. The bot can steer the engagement back to a conversation that happened in the past:

> ***Starwars-bot***: Remember we talked about Yoda a few days ago? Let's get back to that. I can tell you...

I have seen this happen in some bots, such as the SmarterChild bot on AOL (bought and decommissioned by Microsoft). This type of recollection might be surprising and delightful to the user.

Closing Thoughts

Managing context and memory is probably one of the hardest aspects of designing bots. It is also an area of growth from a technology point of view. Looking at logs and understanding when and where your bot loses context or gets amnesia is a good start on the path to mitigating it and designing conversations that are more efficient and memorable.

Building a great conversation workflow is a good start, but as a designer you also need to think about getting the bot in front of your end users. In the next chapter, we'll explore different distribution methods and their implications on your bot's design.

[11]

Bot Discovery and Installation

The greatest obstacle to discovery is not ignorance—it is the illusion of knowledge.

—**DANIEL BOORSTIN**

IN THE PAST FEW years, discovery has been one of the top challenges for software. Users are flooded with websites and mobile apps that offer to make their lives happier, more interesting, and more productive. Advertisers and marketers, growth hackers, and product managers are all trying to solve the problem and get their services to the right users at the right time.

On the one hand, bots have it a little better these days. The ecosystem is still young and not overly saturated. If you go to any of the bot directories, you'll find that they are still considerably less crowded than the app stores. On the other hand, a lot of the discovery mechanisms that are available to mobile app developers have yet to mature for bot developers—a/b testing, analytical tools, and ads are all in their infancy when it comes to bots.

Let's explore a few of the bot discovery options that are currently available.

Bot Directories

Bot directories are websites or in-chat product areas where you can search for, get information on, and install bots. Most messaging platforms provide bot directories in some shape or form. There are also directories, such as Botlist, that are external to the platforms and provide a third-party listing of bots on multiple platforms.

Some bot directories, like Slack's, have featured placements that are curated by a human and rotated periodically, as well as ordered category listings (see Figure 11-1).

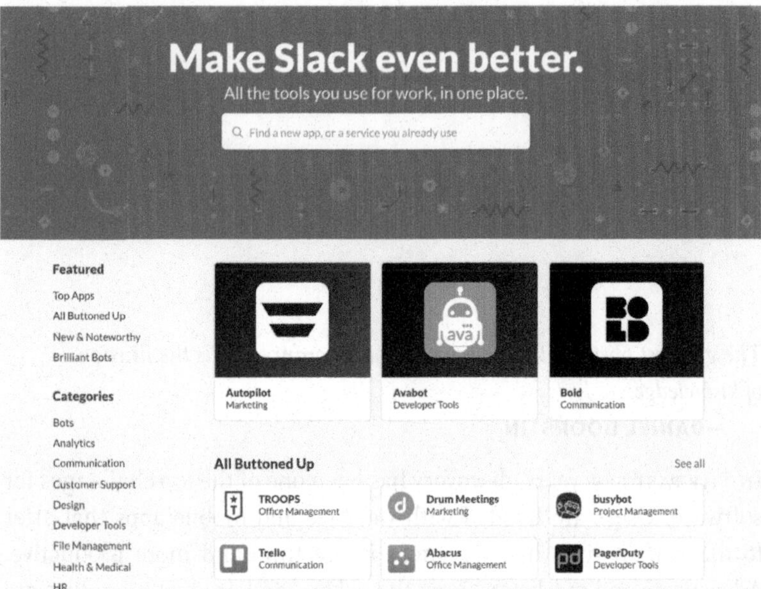

FIGURE 11-1.
Slack's bot directory offers featured picks and regular category listings

The bots featured in the three central slots change every two weeks. Bot owners typically report a factor of 10 increase in users in the time period when their bot is featured.

While Slack has a dedicated directory, Facebook Messenger exposes the directory embedded in the client, alongside featuring slots and search capabilities (Figure 11-2).

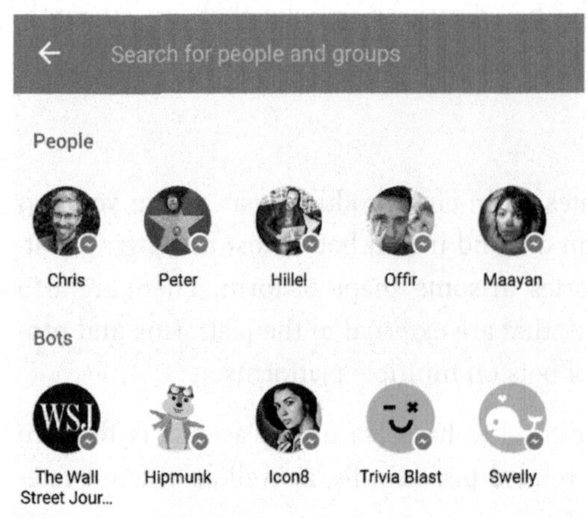

FIGURE 11-2.
Messenger lists featured bots alongside the user's friends, and offers a search capability

Facebook consolidates chats with a user's Facebook friends and bots into a single experience. Messenger does not provide a directory service at the moment, but I am not sure it's needed with this approach (and the fact that it's a consumer platform).

Kik provides users with a very similar in-client experience (Figure 11-3).

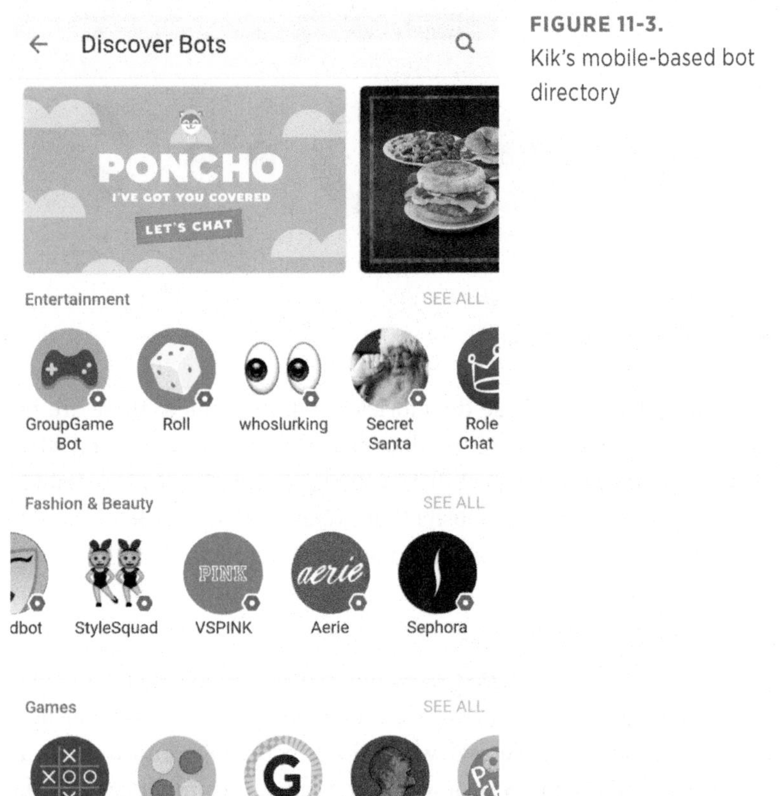

FIGURE 11-3. Kik's mobile-based bot directory

It also has a website similar to the Slack directory for web-based bot discovery (Figure 11-4).

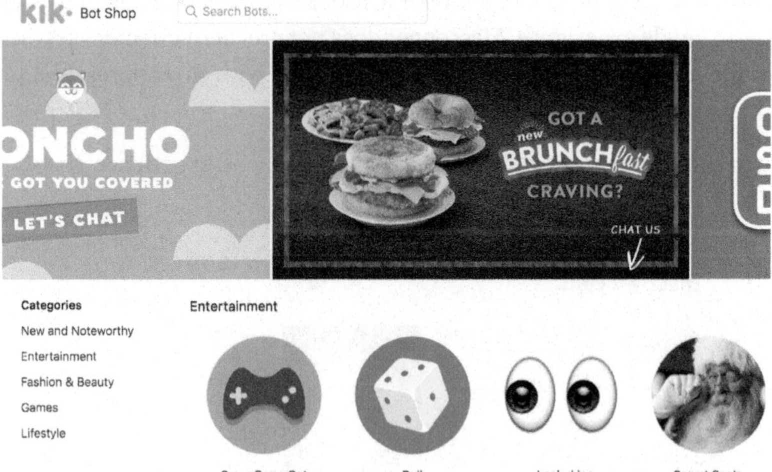

FIGURE 11-4.
Kik's web-based bot directory

The listing details in these directories are quite limited; there are no user reviews, stats, images, or videos that help users understand what the bots are good for. I expect that to change in the near future, but for now, designing your own website that explains what your bot is about, with videos or images of the bot in use, is highly advised.

App Review Process

Most directories have an app review process for the bots that are listed. Details of these review processes are published on the developer portal; they typically include reviews of the functionality and user feedback, as well as a security review in some cases. It is important to note that these review processes are manual and might take more time than expected. Make sure you've adhered to the guidelines provided on the platforms' developer sites before submitting your bot, in order to save a lot of time and additional review cycles.

> **[KEY TAKEAWAY]**
>
> Bot directory review processes are manual and might take more time than expected.

Direct Installation Links

Direct installation links are embedded URLs that direct the user to install or connect with your bot from a website or bot platform. Direct installation links are a great way to share information about bots and promote their installation on the web.

Facebook provides a short link that can drive users to your bot. The URL follows the following pattern: *https://m.me/PAGE_USERNAME*. You can also expose this URL via a web widget Facebook provides (Figure 11-5).

FIGURE 11-5. The "Send to Messenger" widget

Clicking on the button will lead users directly to a conversation with the bot, linked to the widget.

Similarly, Slack provides an "Add to Slack" button that can be shared on your page (Figure 11-6).

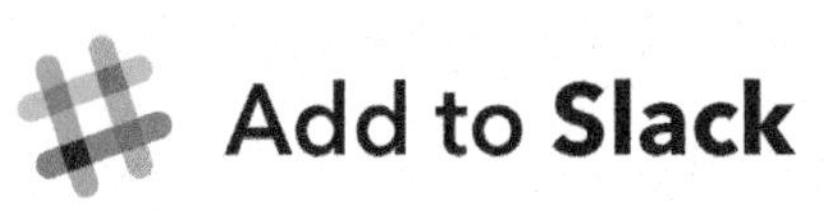

FIGURE 11-6. The "Add to Slack" button

Slack is a platform for business bots, and as such requires the installer to grant permissions to the bot. For example, some bots will require permission to list all the users in a team, while others might not. Clicking on the "Add to Slack" button initiates an OAuth process that takes the user through approving the permissions requested by the bot. Depending on your bot's design, you will need to pick the right permissions to request from the user. Note that asking for more permissions than you need might make a user suspicious of your bot and less inclined to install it. The available permission scopes and actions are listed on Slack's developer portal (*https://api.slack.com*).

Using links and web widgets is a good way to drive users to your bot from existing web real estate (i.e., a web page you already own). Let's say you have a website for your service that gets a lot of traffic—driving that traffic to your bot provides your service with another touchpoint with your audience that was not available until now.

QR Codes

A QR code, in this context, is an image that encodes the bot's identification. By scanning the code using your phone's camera, you can install the bot in your chat application.

As Facebook Messenger and Kik, to a greater extent, focus on their mobile clients for bot distribution, these platforms both enable the use of QR codes for bot discovery. Figure 11-7 shows a few examples of QR codes, one for the CNN bot on Facebook Messenger and the other for the H&M bot on Kik.

FIGURE 11-7.
QR codes for the CNN (Messenger) and H&M (Kik) bots

You can actually scan these and install the bots straight from this book. The key value of QR codes is that they can be placed in the physical world. This means that brands and consumer services can promote the installation of their bots by putting these codes in their places of business or including them with promotional giveaways. This is very common in customer use cases, especially in Asia and with young audiences all around the world.

@Mentions

@Mentions usually refer to mentioning someone in a chat conversation. The @mention typically triggers some sort of notification in the mentioned user's client, and also creates a link to the mentioned user or bot. Figure 11-8 shows how it looks in Slack.

FIGURE 11-8. @Mentioning a bot in Slack

When a user @mentions a bot in Kik, something interesting happens. The @mention not only sends the message to the mentioned bot and lets the bot answer directly in the conversation, but also lets other members in the conversation connect to the bot by clicking on that @mention. In Figure 11-9, I mentioned the @urbanbook bot in a chat with Ben and added the bot to the conversation.

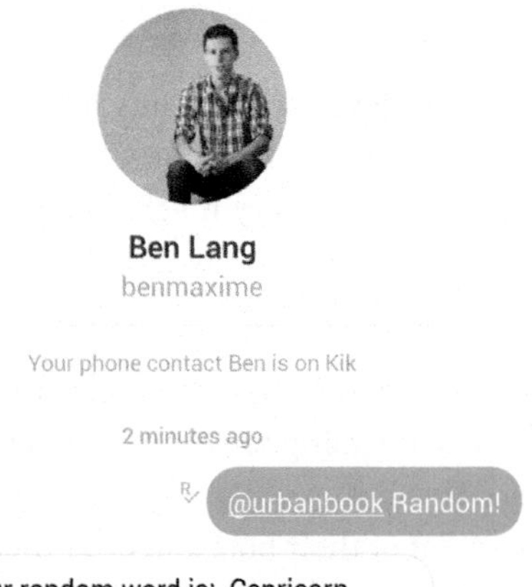

Ben Lang
benmaxime
Your phone contact Ben is on Kik
2 minutes ago
@urbanbook Random!
Your random word is: Capricorn
A sign of the zodiac that involves anyone whose birthday falls between December 22nd - January 20th. The representative animal is the goat.

FIGURE 11-9. Adding a bot to a conversation with a friend is a good way of connecting them in Kik

This behavior creates an interesting contextual viral engagement: a friend mentioning a bot creates a strong incentive to connect to the bot and start a conversation with it.

Bot Referrals

Referring the user to another bot when your bot cannot address the user's intent is an interesting use case that is discussed a lot in the market, although I have seen very few actual implementations of it in real life (we saw one example in Chapter 8). Bots can potentially refer users to other bots within a conversation:

> *User*: And I will also need a ride from the airport to the hotel.
>
> ***Travel-bot***: Well, I cannot help you book a ride, but you can talk to my colleague LyftBot and we can sort this out. Here is the link to install the LyftBot - [link].
>
> *User*: Fantastic, installing now!

The potential for bot-to-bot referral is huge. As you can see in this conversation, the bot captured the intent of the user to perform an action that it could not deliver. Delegating that intent to another bot makes the Travel-bot more useful to the user compared to the alternative, which is to leave the user with a dead end.

Closing Thoughts

Each of the discovery mechanisms described here is valid, and you can choose several of them to ensure potential users have multiple ways of discovering your bot. Each of these discovery mechanisms captures a user's intent in a different way and exposes it with a different level of detail.

In years to come we are going to see more and more discovery mechanisms. Bot directories and stores are going to become bigger and more sophisticated, and developers and designers are going to have better tools available to capture the intent of their bots and offer them as solutions to users' needs.

Once users discoverer and connect to your bot, you'll need to pass the next challenge—next, we'll dive into the different techniques for keeping your users happy and engaged.

[12]

Engagement Methods

Gravitation is not responsible for people falling in love.

—ALBERT EINSTEIN

One of the vanity metrics of the mobile app ecosystem is installs of apps versus uninstalls of the same apps. While this is a suboptimal method of measuring success in the mobile world, mainly because it does not tie in to business objectives or to engagement, this metric is completely useless in the bot ecosystem. Most bots do not get uninstalled; they get abandoned, forgotten... sad.

Dr. Jacob Greenshpan presented a theory in which he compared couples' relationships and our relationships with mobile apps:

1. You install the app—not sure if it is the right fit for you.
2. You love it—it is amazing and it has no faults.
3. You become proficient with it—you come to learn the good and bad in the app.
4. You grow tired of it—the faults are growing, or boredom kicks in.
5. You install another app and fall in love with it.
6. You uninstall the original app.

While this might be an extreme analogy, it usually generates a lot of nodding when I mention it in lectures. The analogy might even be more accurate when it comes to bots. It is hard to stay emotionally detached from all but the most utilitarian bots, and most people use strong feelings and words when they describe bots.

Dr. Greenshpan had a few tips for a good app–user relationship that can apply to bots:

1. Create a great first impression—make the other side fall in love.
2. Keep on adding value—value wears off.
3. Continue evolving your design—users like to be slightly and pleasantly surprised from time to time.

First Impression

The first feeling most people have with regard to products and services they encounter is often either indifference or suspicion. We do not know and therefore do not care about them, or we are not sure if they will actually be good for us. We are constantly being bombarded by services and products that seek to engage us physically, mentally, or emotionally, and it can be hard to figure out which are worthy of our attention.

There are a lot of aspects that make for a positive and actionable first engagement. One key to a positive first engagement is a clear *intent*—users must need or want the service or product, whether they are aware of it or not. Another key is *product fit*—the ability of the product or service to address that intent or need. Then there needs to be a moment of *serendipity*—a moment where the users understand the value of the product or service and realize it is beneficial for them. Then comes the hard work of keeping the fire going and building a *habit*. Demonstrating value and creating a habit should be part of your bot's onboarding experience.

[KEY TAKEAWAY]

Demonstrating value and creating a habit should be part of your bot's onboarding experience.

We talked about installation and discovery in the last chapter, so let's assume the first engagement the user has with your bot is after the installation.

From an engagement perspective, the onboarding needs to achieve several things:

- Clearly define the purpose of the bot and the *intent* it is solving.
- Educate or inspire the user about the *product fit* of the bot and how it addresses the intent.
- Generate a trigger that will build a usage *habit*.

An interesting way to approach bot engagement is through a process described by Nir Eyal in his book *Hooked: How to Build Habit-Forming Products* (Portfolio). Nir describes four steps in the habit-forming process (Figure 12-1):

1. *Trigger*—An internal trigger (such as boredom, anxiety, or curiosity) that is cued by an external trigger (such as a notification or direct message from the bot) and drives the user to take action using the product.
2. *Action*—A simple action you take that yields a reward.
3. *Reward*—The realization of the value: scratching the itch or addressing the intent.
4. *Investment*—An action that makes the service better with use and generates future triggering opportunities.

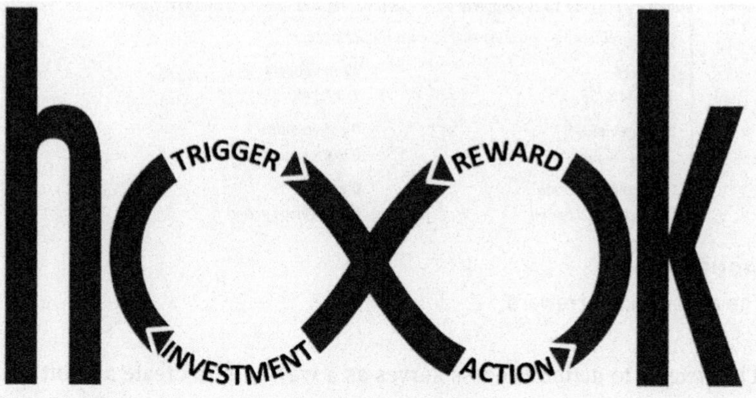

FIGURE 12-1.
The Hook model

You can read about the Hook model in depth at *NirAndFar.com* and in Nir's book, but for now let's explore how this can be relevant to bots. We will use Statsbot as an example.

After the user installs Statsbot into a Slack team, the bot sends this message in a DM to the user (Figure 12-2).

statsbot BOT 8:25 PM
Greetings @amirshevat!
Thanks for adding me to Stryx. Say `help` to get started or `summary` to get today's metrics.
I'm even more powerful when I'm invited to one of your Slack channels. Inside the channel, type `/invite @statsbot`

FIGURE 12-2.
Greetings from Statsbot (the trigger)

The *trigger*, in this example, is the need (curiosity, excitement) to receive daily metrics from the user's Google Analytics service. Users install Statsbot to stay up-to-date—while Statsbot provides many ways to slice and dice the data coming from Google Analytics, the most generic and common need is to get a summary of the stats.

As you can see, Statsbot immediately suggests that the user take *action* and run this basic query (Figure 12-3).

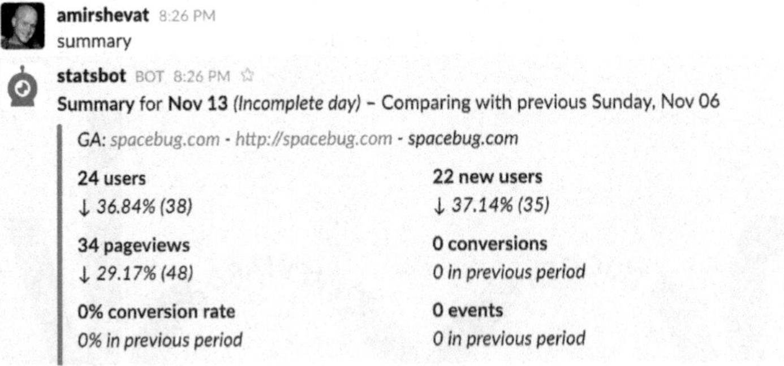

amirshevat 8:26 PM
summary
statsbot BOT 8:26 PM
Summary for Nov 13 (Incomplete day) – Comparing with previous Sunday, Nov 06
GA: spacebug.com - http://spacebug.com - spacebug.com
24 users
↓ 36.84% (38)
22 new users
↓ 37.14% (35)
34 pageviews
↓ 29.17% (48)
0 conversions
0 in previous period
0% conversion rate
0% in previous period
0 events
0 in previous period

FIGURE 12-3.
The action and the reward

This *trigger* to *action* process serves as a way both to create a habit and to educate the user about how to use the bot. The user in this example follows the bot's prompt and takes action, and gets an immediate *reward* in the form of a report. Talking to Nir about this example, he also mentioned that this is an example of a "variable reward"—the fact that the stats are always changing has a slot machine–like effect that, used correctly, can serve to bring the user back time after time.

The user now understands how to communicate with the bot and has realized the basic value of the bot. Now it is time to move to the *investment* stage (Figure 12-4).

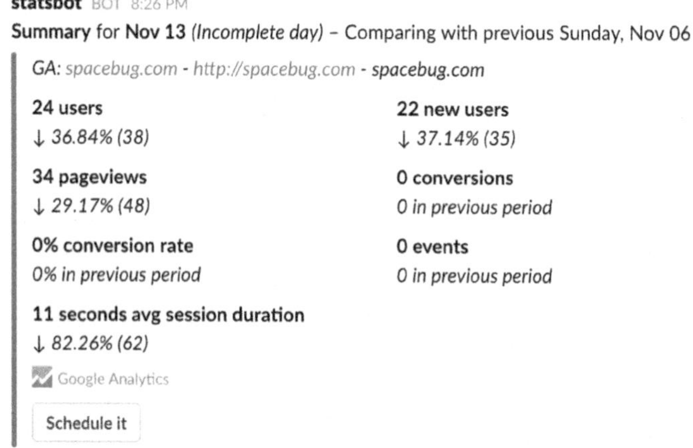

FIGURE 12-4.
The investment (scheduling)

Note the button at the bottom of the report—this is a call to action to build an ongoing habit. Clicking on the "Schedule it" button prompts the user to invest a few moments in setting up a scheduled notification with the report (Figure 12-5).

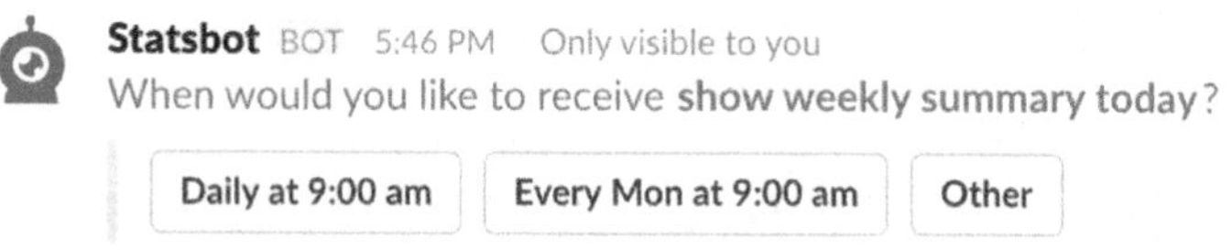

Statsbot BOT 5:46 PM Only visible to you
When would you like to receive show weekly summary today?
Daily at 9:00 am
Every Mon at 9:00 am
Other

FIGURE 12-5.
Statsbot scheduling a report

And now we have a habit formed and set by the user. When delivering reports, Statsbot attempts to create more habits by prompting the users to schedule other reports and encouraging them to create more sophisticated queries (Figure 12-6).

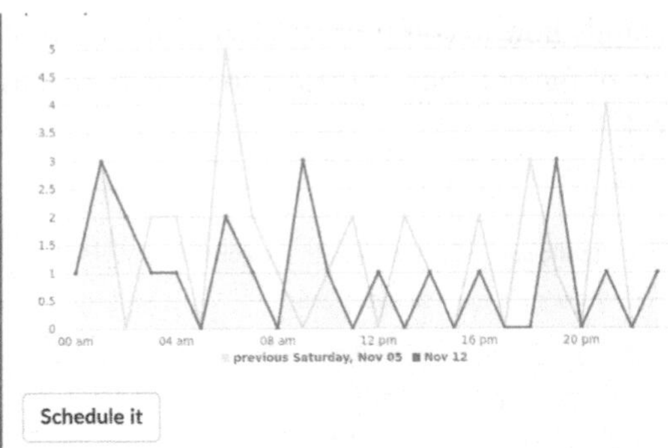

I can also segment metrics by dimensions. Ex: `@statsbot users today by country`

FIGURE 12-6.
Statsbot continuously educates and promotes habit building with the user

Suggesting different and advanced ways to engage with the bot is a very useful pattern in this case, because it keeps the bot interesting to the user.

The Statsbot team ran into this best practice by mistake—they saw that most users were not using all the features they had built, and limiting themselves to a small subset of the functionality. Additionally, there was little usage of new features and queries, because users just did not know about these new features. The team started experimenting with hinting to the users that they could do more with the bot, and were pleasantly surprised to find that users loved that feature and that it led to higher levels of user engagement.

If we look at Poncho as a consumer use case, we see the same pattern. The bot starts by introducing itself and identifying the *trigger*—the need to know the weather forecast (Figure 12-7).

Oh, hey, Amir! I'm Poncho and I'm here to talk about weather. You ready?!

Let's do this thing!

I can send you daily weather forecasts! Where do you live? Tell me the name of your city, neighborhood, or postal code.

FIGURE 12-7.
Poncho introducing itself (the trigger)

Then it prompts the user to take *action* and set their preferred location. Following the user's action, the bot provides an immediate *reward* in the form of a weather update (Figure 12-8).

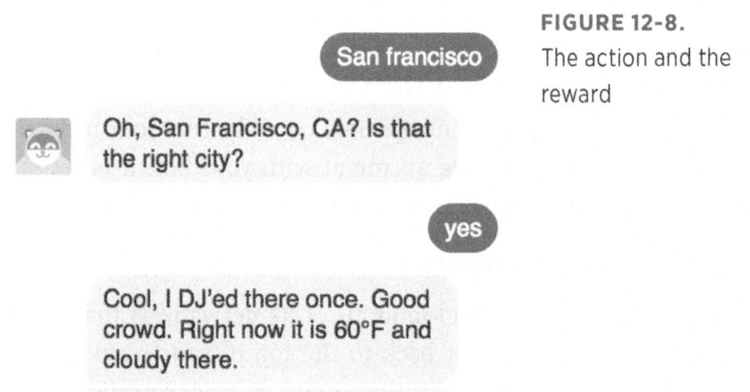

FIGURE 12-8. The action and the reward

Next, the bot asks the user to invest some time in setting up a recurring update (Figure 12-9).

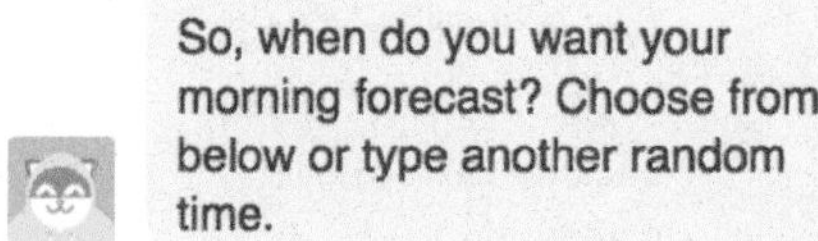

So, when do you want your morning forecast? Choose from below or type another random time.

FIGURE 12-9. The investment

Again, this initial interaction forms a habit: getting morning forecasts from the bot. After this habit has been set, users do not need to do any additional investment and move to a usage pattern of automatically getting their weather forecast every day from Poncho.

In both examples we witnessed the pattern of creating a trigger, promoting an action, providing a reward, and asking for an investment that ends with users opting into a habit-forming engagement with the bot.

One of the key differences between this and just automatically setting a notification from the bot is that in these examples the user has consciously opted to receive the periodic notifications. The act of opting in makes users less inclined to consider these notifications as spam and more likely to interact with them.

Onboarding is a great place to get users to input details (such as location and other preferences), set the cadence for repeatable actions, introduce the bot to the team, and generate additional interaction. Take advantage of this stage to create a positive first experience.

Ongoing Engagement Points

One of the reasons for lack of engagement could be that your users do not have an easy way to initiate an intent with your bot. It is recommended that you support several engagement points to your service.

Here are a few ways you can do that:

- Support a "help" command (and all of its derivatives) that users can use at any time to get back to the top of your service navigation flow (unless you are in the context of a conversation; then the help should be context-aware). Figure 12-10 shows how Google Assistant does this.

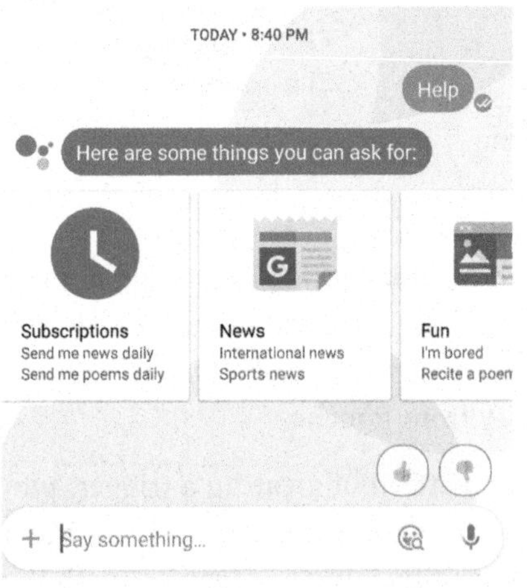

FIGURE 12-10. Typing "help" in Google Assistant takes the user back to the top-level navigation

- Support "hello" (and all of its derivatives) as a way to start a conversation. It sounds trivial, but this is not supported in many of the bots I have tested.
- Explore additional platform-specific entry points. For example:
  - Slash commands in Slack
  - Persistent Menus in Facebook Messenger
  - @Mentions in Kik and Slack

NOTIFICATIONS

Another common engagement point is sending notifications to your users. Most platforms let bots send users notifications in one way or another. The risk with unsolicited notifications is that they can be perceived as spammy and might motivate users to disengage from or even uninstall or block your bot. If you are using notifications to drive engagement, make sure you capture the user's intent to receive them, and ask for permission before sending them ongoing notifications. Make sure your notifications are valuable and, when applicable, make sure you provide the user with a way to opt out from unwanted notifications.

[KEY TAKEAWAY]

Make sure your notifications are valuable and give your users a way to opt out.

Here is an interesting insight from Dmitry Dumik, CEO of Chatfuel, a popular bot-building tool for Facebook Messenger:

> One of the biggest opportunities that chatbots possess is an ability to deliver a message right to a user's inbox in the form of a push notification. At the same time this is one of the biggest threats since there is no way to reach back as soon as a user blocks you. After analyzing billions of messages sent through our platform we found out that the rate of unsubscribes skyrockets if a user gets more than 1.9 messages/day. That doesn't mean you should send ~2 messages a day, but you definitely should have a very good reason to send more than that.

PROMOTING ENGAGEMENT BY JUST BEING USEFUL

When I analyze the best engagement I have with bots, I realize that the same rule that applies for mobile apps applies for bots—be useful, address a pain, and users will come back to you again and again. Ultimately, engagement cannot be artificially generated; it stems from your bot's usefulness.

In my house, we wake up every morning and ask Alexa about the weather. I CC my scheduling bot, Amy, on all meeting requests because I do not want to have to coordinate my meetings myself. I use the Slack

CRM integration because it provides context in a chat conversation. These bots do not need to regularly seek to recapture my attention in order to be useful; they are important for me to engage with because they add value to my life and my work.

[KEY TAKEAWAY]

Engagement cannot be artificially generated; it stems from your bot's usefulness.

Closing Thoughts

When external triggers, like notifications from the bot, are aligned with the internal trigger and gratification of getting the users what they need, when they need it, that is where the magic happens and your bot becomes priceless. It is always hard to build great value—but if it were easy, everyone would be doing it already.

Next we will explore the challenge of making money from your bot. We will see examples of bots that are already monetizing in different ways and learn from their designers' experience.

[13]

Monetization

A wise person should have money in their head, but not in their heart.

—JONATHAN SWIFT

WHY ARE WE TALKING about making money in a design book? Well, many aspects of business are tightly connected to design choices. Understanding how your bot is going to drive revenue is an important part of understanding your engagement with your users.

Let's get one thing clear: a bot is a type of user experience, and a way to expose products, services, or a brand. The only way to make money out of bots, without having a service or a product, is to be a bot builder and have someone pay you to build that bot. End users do not pay for bots, they pay for the services the bots expose; they pay for the products they promote and they are influenced to connect to the brands they represent. But still, in the same way we now have companies that call themselves mobile businesses because they make most of their revenue through mobile, we are going to see more and more companies calling themselves bot companies.

[KEY TAKEAWAY]

End users do not pay for bots, they pay for the services the bots expose.

In this chapter we will review the direct and indirect ways that bots can drive revenue. This is not a comprehensive list, as I am sure entrepreneurs will come up with many other ways to make money in this industry.

Subscription

This is currently the most common way that bots drive revenue—the bot provides an ongoing service that the user subscribes to and pays for.

An example of a subscription-based bot is Growbot. Growbot is a bot that helps teams celebrate personal and professional wins; it promotes great team spirit and cohesiveness. The bot looks for keywords like *kudos* and collects and amplifies these congratulations (Figure 13-1).

Major kudos @joseph on the new splash page.
We look like a real company now!

growbot BOT
You're kind of a big deal @joseph 138

FIGURE 13-1.
The Growbot HR bot

Most of the features provided by Growbot are free—the company does not charge to install the bot and to use its core features. The team deliberately did not want to charge for aspects of the bot that promote usage, engagement, and virality. They did, however, carve out a set of features that are important to big companies and HR departments. Growbot charges for a premium service that includes company leaderboards and keyword customization.

Growbot asks users to pick a plan during setup. Figure 13-2 shows how the freemium-based Growbot highlights the premium functionality on the web.

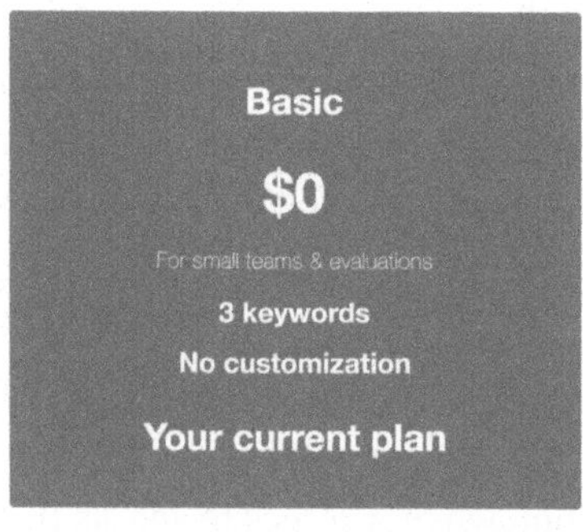

Upgrade Growbot to fuel insights and promote your company values
Growbot Premium includes access to premium dashboards, employee leaderboards, company values support, and more.

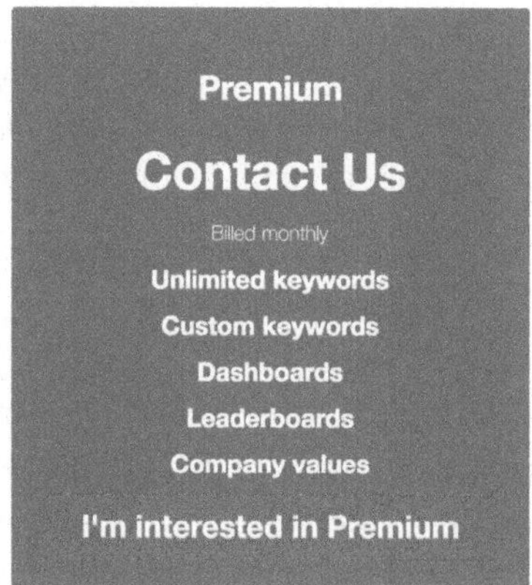

Company values

FIGURE 13-2.
Growbot experimenting with premium services

Another example of a subscription-based bot is Statsbot, an analytics bot that exposes stats from Google Analytics and Mixpanel. Statsbot's developers decided to go with a trial model in which they provide you with a fully functional premium bot for a trial period. An interesting strategy that Statsbot applies is to let the bot itself prompt the user for payment (Figure 13-3).

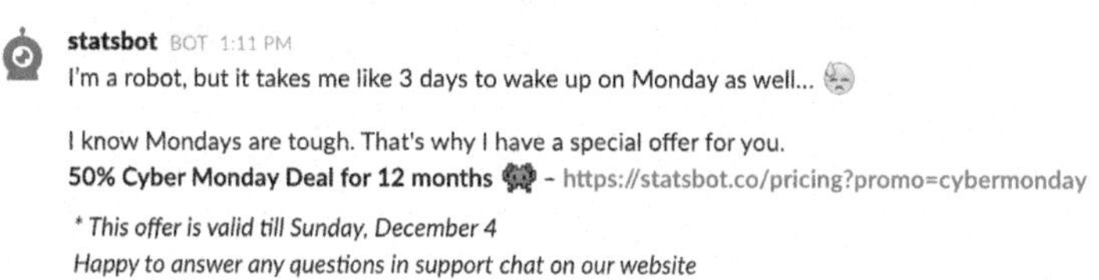

statsbot BOT 1:11 PM
I'm a robot, but it takes me like 3 days to wake up on Monday as well...
I know Mondays are tough. That's why I have a special offer for you.
50% Cyber Monday Deal for 12 months - https://statsbot.co/pricing?promo=cybermonday
* This offer is valid till Sunday, December 4
Happy to answer any questions in support chat on our website

FIGURE 13-3.
Statsbot prompting the user to start a subscription

As you can see, both bots drive the user to set up the subscription on the web—this is because none of the chat platforms currently support in-platform payments. Driving the user out of the conversation and to a website might impact the conversation, but currently that is the only option available for subscription bots.

Note that even though the payment itself is done on the web, it is important to acknowledge it in the chat interface. Notifying the user of upgrades, downgrades, and recurring payments is a good way to promote transparency and user satisfaction.

The subscription model, when done right, is very lucrative because the lifetime value of each user (the amount of money the user will pay in the lifetime of using your product) is usually higher than with other models.

Ad Serving

Ad serving has been the bread and butter of many web and mobile businesses. Bots are in a unique position when it comes to ads, as they can build a personal relationship with the user, as well as collecting a lot of personal information that can contribute to more targeted and finely tuned ads that lead to a better click-through rate.

In the example in Figure 13-4, the user actually asks the bot (a teen-focused "influencer bot" on Kik) for the ad, in the form of a coupon, as part of the conversation. Most advertisers will tell you that a user asking for an ad is as good as it gets when it comes to conversion to paying users. The bot captures the intent at exactly the right moment and serves a relevant ad to the user.

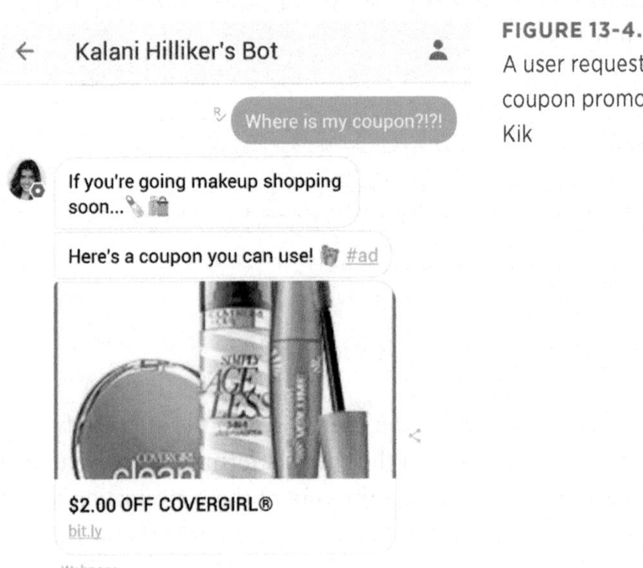

FIGURE 13-4. A user requesting a coupon promotion on Kik

I talked to Andy Mauro, who built this bot, and he stressed that the motivation was to provide an engaging fan experience as well as to deliver value to the brand, and to do so in an authentic way that is more conversational and personal than traditional advertising.

Note that while some platforms are happy with bots serving ads, other platforms do not allow it. Slack, for example, does not permit the serving of ads on its platform. The reason for this is that Slack positions itself as a business communication platform; it would not make sense to serve ads in a business tool context that promotes productivity and focus. If you are thinking of ads as the business model for your bot, consumer-facing platforms like Kik can be great venues for your bot.

Data—Analytics and Market Research

Bots can collect a lot of data about user preferences and interests, through engaging in a conversation or playing a game.

An example is the Swelly bot. Swelly lets the user play a game of choosing between two options, and also lets users and marketers ask questions to get help with decision making. After choosing an option, the user is rewarded by being shown what other users chose (creating a social hook to reengage with the bot) and prompted to make another choice. Figure 13-5 shows what it looks like on Facebook Messenger.

FIGURE 13-5.
The Swelly bot

Now imagine you were a fast food brand, for example, and wanted to know which new type of sandwich customers would prefer, or what type of fast food imagery they would find more appealing. This type of bot would be a gold mine for your research. The Swelly bot can generate detailed reports on audience preferences in a precise and very fast way. While this bot does not charge the end user, I am sure it will make a lot of money helping brands understand their audience.

Selling Goods and Services

Bots can also become the channel through which you sell goods and services. These can be tangible goods that the bot sells directly in chat, or paid services that the bot exposes through chat.

An interesting example of a paid service exposed via chat is the ride services Lyft (Figure 13-6) and Uber (Figure 13-7), which have developed interfaces for both Slack and Facebook Messenger.

Lyft BOT 10:41 AM Only visible to you
/lyft eta 155 5th SF
Pickup: 155 5th Avenue, SF, CA, United States
Lyft Line ETA is 3 min
Lyft ETA is 3 min
Lyft Plus ETA is 4 min
+
Message #general

FIGURE 13-6.
The Lyft bot on Slack

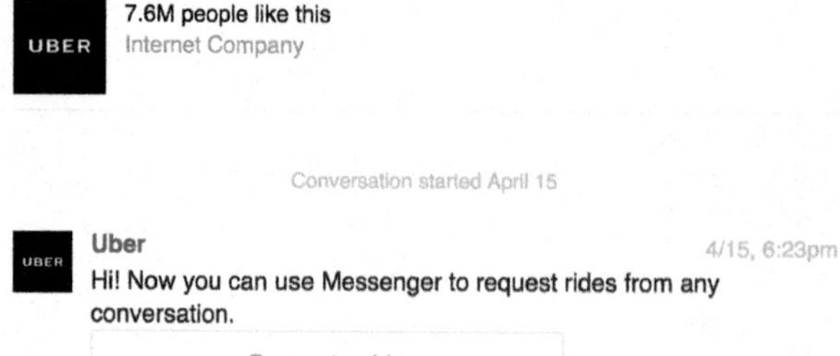
UBER
7.6M people like this
Internet Company
Conversation started April 15
UBER
Uber
4/15, 6:23pm
Hi! Now you can use Messenger to request rides from any conversation.
Request a ride

FIGURE 13-7.
The Uber bot on Messenger

I find these examples very interesting because these are services that were previously available on mobile and consumed mainly in the form of a mobile app. The conversational interface lets you consume these services and order a ride without having the app installed on your phone. As the bot market grows, we might see more and more paid services move to the chat platforms.

We're also seeing goods sold via bots—and we cannot talk about this without mentioning the Amazon Echo. Since we introduced Alexa to our house and started using it in our kitchen, more and more of our household goods are being ordered through it. "Alexa, add napkins to the shopping list" is easy and frictionless; the intent is captured at the right time, by a bot that can address that intent. Amazon has also launched Alexa's "deals of the day," which are deals only available through the bot.

Referral Fees

This is another major business model on mobile and web that is moving to bots—the bot can help you decide what to buy or which service to consume and then refer you to the right service, rather than actually completing the transaction itself.

A good example of a referral business model in bots is the Kip bot, available on Slack, Messenger, Kik, and Telegram. Kip is a shopping assistant for teams; as well as coordinating purchases across the team it helps you find items to buy and then refers you to the merchant that can fulfill the order (Figure 13-8).

The advantage of Kip is that it is not bound to a single vendor; it can provide the user with results sourced from multiple merchants. When the user completes a transaction on a merchant's site, Kip gets a referral fee for sending the user to that vendor.

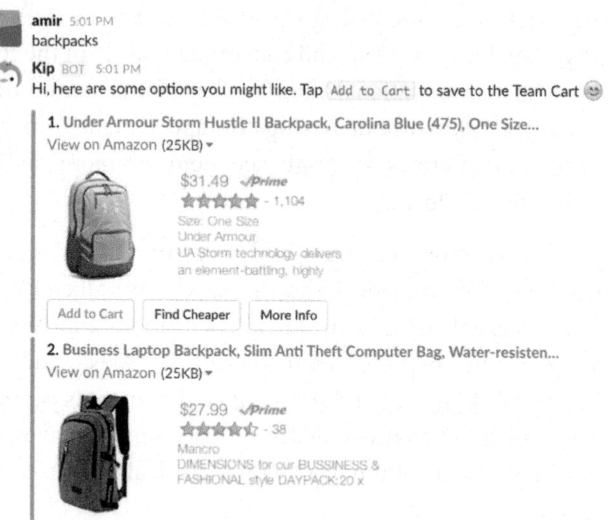

FIGURE 13-8.
Kip bot referrals

There are many other use cases in our lives that follow the same pattern, from travel agents to car salespeople, and we will likely be seeing a lot of agents like Kip in the future that will help us choose between vendors.

Brand Promotion

Another indirect way to drive your business through a bot is by promoting brand recognition. Thinking about the bot as a frontend representative of your product or service can make a lot of sense, and having a delightful bot that provides a useful service can generate strong attachment to your brand.

This pattern is being explored by many marketing managers. I think the key, however, is to focus on the value proposition the bot can provide and not only its persona—both of which will reflect heavily on the brand perception customers will develop while engaging with the bot.

Figure 13-9 shows an example of a branded bot on Kik.

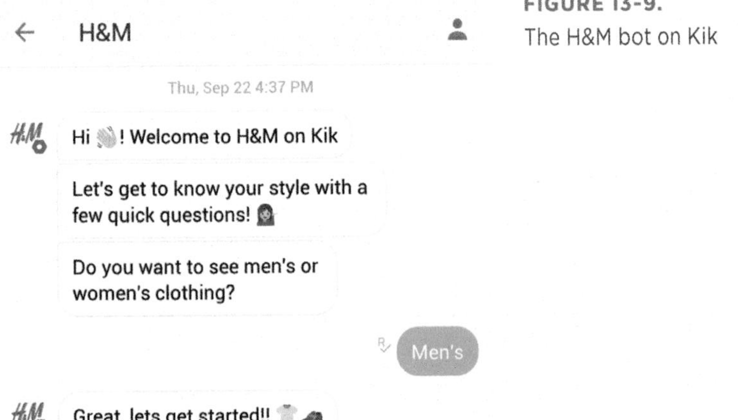

FIGURE 13-9.
The H&M bot on Kik

As you can see, the bot clearly represents the H&M brand. It uses a young and friendly persona (the use of emojis and casual manner), but also puts an emphasis on providing value to the user—in this case, helping them search for clothes—very early in the conversation. Providing value is key because a bot that does not do that might actually generate a negative brand association, similar to a company representative who keeps wasting your time.

EXTENDING A PAID-FOR PRODUCT

Extending a paid-for product or service is a very common pattern for companies that already have an established offering and want to extend this offering to another user experience. We have already seen this pattern with the introduction of mobile, with companies like Concur developing mobile apps that extended their services and let users report expenses while on the go.

Trello is a well-established paid service that provides help with task and project management. Noticing that a lot of their clients were using Slack as a way to communicate, the team at Trello decided to extend their service to that new platform and provide parts of the functionality inline in the chat interface (Figure 13-10).

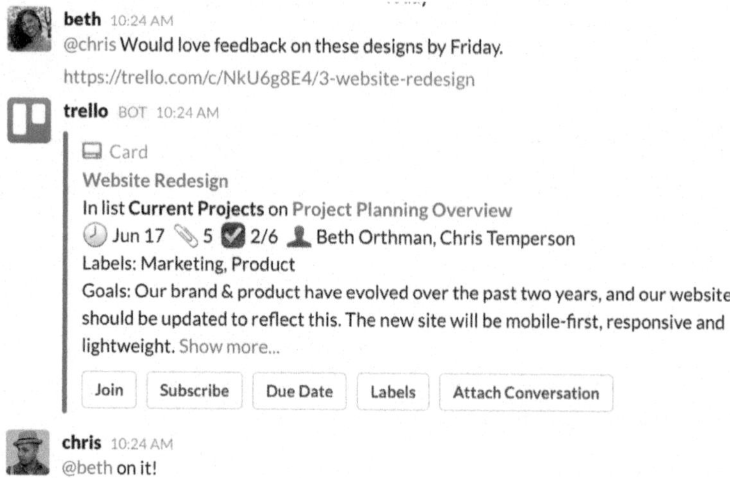

FIGURE 13-10.
Trello on Slack

As you can see, the interaction is contextual and actionable in the context of the conversation. It makes a lot of sense for users to interact with the service in this intuitive and collaborative way.

Choosing which part of the functionality to expose in chat is key to the success of use cases like this. If you expose too much functionality, you might provide a cumbersome experience, and if you provide too little it might be not valuable enough.

Why is extending to new platforms useful to your business model? The first reason is that it improves engagement and the usefulness of your service—users have yet another touchpoint to get value from the service. The second reason is that it improves your product's defensibility, by mitigating the risk of a competing product gaining traction on this new platform and stealing your clients.

IN-BOT VIRTUAL GOODS

Sale of in-bot virtual goods is another model that I think is going to be big in upcoming years. An example would be a game bot that sells you chips or power-ups, or a bot that sells you music and photos. While I could not find good examples of in-bot payment in real life today, this model has been one of the most lucrative models in the mobile app industry.

When Should You Start Charging Users?

I interviewed several bot builders, and they all came up with the same answer: you should only start considering charging for your bot when you feel that you have reached product/market fit. Users need to get hooked on your product, use it on a regular basis, and derive a lot of value from it before they are asked to pay for the bot. Whether you are providing a trial, a freemium version, or in-bot virtual goods, make sure your users love your product before you try to charge them.

> **[KEY TAKEAWAY]**
>
> You should only start considering charging for your bot when you feel that you have reached product/market fit.

Another point many bot providers have stressed is that it is important to decide what to charge for. Charging for features that promote engagement or the virality of your product might be counterproductive.

Closing Thoughts

When I joined Google eight years ago, we all thought that paid apps were going to be the way developers would make money out of mobile. What happened in reality is that in-app purchases became the way users preferred to spend their money. The point is that this industry is very young, and we are still exploring the right monetization strategy. You should explore these and other strategies for your particular use case and experiment until you find what works.

In the following chapters we will take the theory we've learned up to now and put it into practical use. The next chapter describes the steps we need to take to design a bot; then we dive right in.

[14]

Design Process Overview

Design is not just what it looks like and feels like. Design is how it works.

—STEVE JOBS

IN THIS SECTION OF the book, we will be moving from theory to practice. We will explore an actual use case and go through the design process of making this use case a reality. We will also explore different ways to iterate and improve the design once it is done through user feedback.

Talking to several product managers and bot designers has made it clear that there is no single process that works for everyone. I will try to lay out a process that works for many use cases, but you should explore and figure out the best process for you.

The Steps

Here are the core steps that we will demonstrate in the following chapters:

Use case definition and exploration

In Chapter 15 we will describe the purpose and functionality we need to design, and attempt to understand the persona and get clarity on the brand and other attributes that might impact the design. We will then make a basic analysis of decisions based on the use case data, drawing on some early-stage experimentation.

Conversation scripting

In Chapter 16 we will start writing the basic drafts of the scripts for the use case workflows. We will explore where we think we will use rich interaction and rich controls, and where we will use plain text conversation.

Designing and testing
: In Chapter 17 we will demonstrate how to design different aspects of our use case. We will experiment with a few design tools and explore the benefits of each. We will also look at how we can get something simple in front of users to validate some of our scripting and prototyping assumptions.

Bot building overview
: In Chapter 18 we will review different ways to actually implement our bots. While this is a design book, this is an important step, even if you do not plan to actually do the coding.

Analytics and continuous improvement
: In Chapter 19 we will look at what kind of data we can get from our bots once they are in the hands of our users, and demo that with a few bot analytics frameworks. We will show how bot designers have turned this data into insights and adjusted their designs, making their bots better.

The Tools

We will review several tools along the way, but as this ecosystem is experiencing a boom, there are going to be a lot more tools emerging in the near future that will not be covered in this book. For that reason, I will try to stick to design principles rather than tool implementation details. We will not go over all the menu items and shortcuts in all the tools (far from it!); I will, however, describe and demo different aspects of each step in the design process with tools that are available today.

As for the use case, there is no "Hello World!" (super-simple) version of a bot idea, and even if there was, it would not be useful for our exercise. The use case we will be designing is a vacation request system that partners with a travel service to provide bots for business users—users will request paid time off in the business platform bot and get vacation recommendations in the consumer platform bot. We will make note along the way of certain assumptions about technical issues that will be out of scope. Let's dive in!

[15]

Use Case Definition and Exploration

Successful engineering is all about understanding how things break or fail.

—HENRY PETROSKI

Chris Messina really wants to take a vacation to Cancun with his family. Chris works as a DevOps manager in a midsize company in San Francisco. His partner in life has already bought the tickets and booked the hotel, so Chris just has to complete two simple tasks: get approval from his manager, and find fun things to do during the vacation. Here is where our use case begins.

Our task is to build two bots that will help Chris take this vacation:

PTOBot
: The PTO (paid time off) bot will help Chris secure his manager's approval to take a vacation.

VacationBot
: The vacation bot will help Chris find fun things to do while in Cancun.

Both bots are provided by PTO-IT, which is a leading software provider in the field. Both bots share the same infrastructure and will provide Chris with a state-of-the-art seamless experience wherever he is.

The marketing and product departments of PTO-IT have done their research, and have come up with the following insights regarding the bots:

Audience
: Adults aged ~25–55, tech-savvy, tend to shop online, early adopters.

Business model

PTOBot will be an extension of the paid service already provided by PTO-IT, a widely distributed solution with high-tech companies. VacationBot will be a new experiment of PTO-IT in an effort to venture into the consumer landscape and generate revenue from referral fees.

Features

The marketing department recommends that these bots provide a rich and interactive experience, as past experiments with text-only interfaces (through SMS-based approvals) have not been super successful with this audience.

Preferred devices

PTO-IT's engineering department recommends a platform that supports both mobile and desktop devices based on past experience in this industry.

Platforms used

In a set of interviews with samples of the target audience, many of them reported using email and Slack for work, and Facebook Messenger, email, and WhatsApp for communicating with family and friends. Some of them have heard about Kik and SnapChat, through their kids.

For our PTOBot, the following basic requirements have been identified:

- PTOBot will enable users to perform actions in the PTO-IT system (*assumption*: the user's Slack and PTO-IT email addresses will be used to link the accounts—linking the user's Slack account to the legacy PTO-IT system is out of scope).
- PTOBot will enable the user to request PTO, setting dates and providing a description
- PTOBot will enable managers to approve PTO requests (*assumption*: PTO-IT already knows the employee/manager relationships).
- PTOBot will record the approval/rejection status of each request and communicate it to the employee.

- PTOBot will enable an employee's teammates to get notifications when their colleague is taking a vacation.
- PTOBot will promote VacationBot to the user at the end of the approval process.

For our VacationBot, the following basic requirements have been identified:

- VacationBot will only be distributed by PTOBot and work for PTO-IT clients; phase one will not be open to the public.
- VacationBot will get a parameter in the referral from PTOBot, and will be able to retrieve vacation start and end dates from PTO-IT.
- VacationBot will support sending notifications to the user, as well as providing the ability to get a list of events and activities at the vacation destination on demand.
- VacationBot will promote subscribing to notifications about events and activities to the user during the vacation.
- VacationBot will enable linking from each notification to the relevant event or activity with the appropriate affiliation link.

PTO-IT's marketing team insists on keeping on-brand with this new line of service. Here are some brand guidelines:

- PTO-IT has been in the market for the last 15 years—it is a mature and respected brand.
- The brand and products should reflect professionalism, seriousness, security, and high-end quality.
- Interactions and recommended activities should be safe for work (SFW) and family-friendly.
- Productivity is the number one value—get things done, and fast.
- PTO-IT aims to be perceived as innovative and is eager to adopt new platforms—give us bots!

Basic Analysis

Now that we are armed with all this information, we need to make a few decisions before we start designing the bot interactions.

SETTING A PURPOSE

We'll start by distilling the purpose of our bots:

PTOBot

PTOBot enables the process of employees requesting paid time off and managers approving these requests. It also serves as a connection to the system of record (PTO-IT) and improves the transparency of vacations taken by team members.

VacationBot

VacationBot takes leads from PTOBot on users who are planning vacations and provides them with activity recommendations, helping make their vacations delightful.

PICKING A BOT PLATFORM

Let's pick our primary platform for each bot:

PTOBot

This is a bot for a business use case. We heard from our target audience that they use email and Slack for work. We also heard from marketing that they would like to provide rich controls and an interactive experience, which more or less excludes an email bot. There is also a desire to be innovative and adopt a new platform, which is another reason not to use email.

Let's pick **Slack** ;)

Note: Consider adding email support in later versions of the product, as a way to reach non-Slack users (out of scope for this example).

VacationBot

This is a bot for a consumer use case. Our target audience reported using Facebook Messenger, email, and WhatsApp outside of work. WhatsApp currently does not have a bot framework. Email is excluded for the same reasons mentioned above.

Let's pick **Facebook Messenger** ;)

Note: Consider adding email or Kik support in later versions of the product, as a way to reach non-Facebook users (out of scope for this example).

DEFINING A PERSONA

Now let's sketch out the personas for our two bots. First, the PTOBot:

> *Name*: PTOBot
>
> *Environment*: Work
>
> *Audience*: Adults aged ~25–55, early adopters
>
> *Task at hand*: PTO requests
>
> *Runtime variations*: Report abuse, zero tolerance to not safe for work (NSFW) content
>
> *Locally relevant social acceptance*: Work environment
>
> *Service branding*: Professional, productive
>
> *Values*: Getting things done, and fast
>
> *Derived personality*: Serious, to the point, friendly but not humoristic, safe for work. This bot should be like a company office manager or a personal assistant.

Then the VacationBot:

> *Name*: VacationBot
>
> *Environment*: Consumer
>
> *Audience*: Adults aged ~25–55, early adopters
>
> *Task at hand*: Provide fun activity recommendations
>
> *Runtime variations*: Provide a more casual conversation during the vacation itself
>
> *Locally relevant social acceptance*: Family-friendly
>
> *Service branding*: Professional, fun but still not too casual
>
> *Values*: Enriching employees lives while they're on vacation
>
> *Derived personality*: Friendly but not too humoristic, family-friendly, safe for work. This bot should be like the concierge service in a high-end hotel.

CHOOSING A LOGO AND VISUALS

Let's pick a logo and establish our visual preferences. This is not set in stone, as we might learn a few things during the exploration and prototyping phases, but it will give us a basic framework to think about when scripting our bot conversations.

PTO-IT has very strong brand recognition at many high-tech companies, so we will use the company logo for both bots (Figure 15-1). This will ensure consistency and provide the users with the feeling that they are using a PTO-IT solution. Another reason to keep the logo is that marketing wants this to be perceived as an extension of a current offer rather than a new service.

FIGURE 15-1.
The PTO-IT logo

When it comes to color coding, PTO-IT uses a basic red, blue, and yellow schema. Bots should use this color schema when applicable. We will refrain from using funny images and GIFs, as they reflect badly on the serious brand; we will use emojis lightly when applicable.

NAMING CONVENTIONS

PTOBot should use the naming conventions set by PTO-IT (remember to use "paid time off" instead of "vacation," for example).

If implementing slash commands, they should be associated with the parent product. For example:

> /PTO-IT [request]

For our VacationBot, as this is a slightly more casual product, no restrictions on naming conventions apply. When referring to PTO-IT or PTOBot, however, it should keep the "paid time off" naming convention.

Solution Exploration

One of the cool things with this new conversational user interface is that it is easy to explore solutions before you even start designing. This is not even prototyping (we will touch on prototyping in Chapter 17); this is more to explore the conversational paradigm. We will use a method called *Wizard of Oz* user research. The Wizard of Oz technique enables unimplemented technology to be evaluated by using a human to simulate the response of a system—in our case, the bot.

[KEY TAKEAWAY]

With bots it is easy to explore solutions before you even start designing using the Wizard of Oz technique.

In order to explore the vacation bot on the Facebook Messenger platform, I have actually messaged my friend Malvina, who works at Facebook, with a few recommendations for activities for her upcoming trip to Palm Springs. I found a few places we enjoyed on our last trip there, and sent her some recommendations through Facebook Messenger (Figure 15-2).

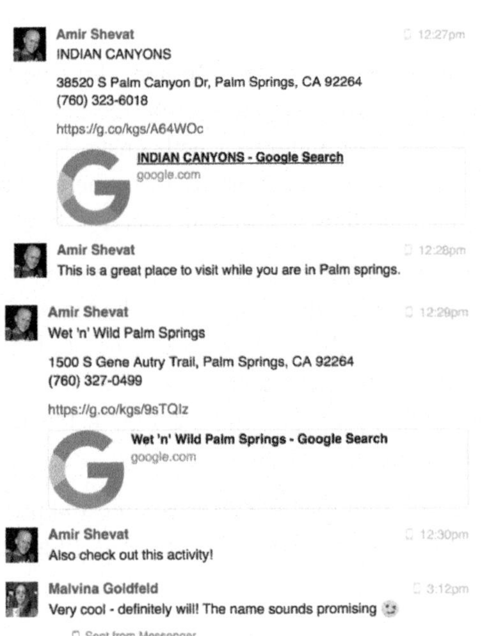

FIGURE 15-2. Exploring the conversational interface for the vacation bot with the Wizard of Oz technique

This is far from the experience I envision for the actual bot—it does not use rich elements like galleries or buttons—but it still reveals a lot of insights with very little effort.

There were a few things that I noticed immediately. First, it felt kind of odd to just post the links; it felt like I needed to add a sentence to add a personal touch to each recommendation.

The second thing I noticed is that it took Malvina about three hours before she saw and reacted to my notifications. This gave me a hint that timing is very important in these scenarios, and that this type of interaction might be more asynchronous than I'd thought. This finding, of course, needs to be verified with more user testing, but it might imply that we should send recommendations the evening before the actual day of an event or activity, to ensure users can see and are able to act on the information we provide them.

The third thing I noticed is that when Malvina answered me, she provided information that might be useful for future recommendations. Understanding that users react well to this type of recommendation and are willing to provide feedback is important. Getting signals from user input is a crucial part of a successful conversational interface, and we should be able to optimize our recommendations to each user from their feedback.

Because faking an account on Facebook is hard, I cheated a little with my Wizard of Oz methodology: I did not pretend to be the bot, but rather mimicked the conversation a bot might have with a user. This is one of the benefits of this conversational user experience, where bots are expected to join the existing paradigm of a human chat interface.

Exploring on the Slack platform is slightly easier because you can just create another user account (or even modify yours) with the logo and name of the bot, and mimic the bot's behavior. This is far from high-fidelity mimicking, as you cannot input rich controls (such as buttons) as a user, but it gives the user with whom you're interacting the feeling that they are not talking to a human.

In one of my work's Slack teams, I modified my display name and profile photo and asked people to send me their PTO requests (Figure 15-3).

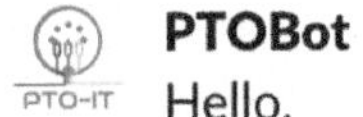

PTOBot 9:18 PM
PTO-IT
Hello,
I am PTOBot - I would be happy to set up your next PTO.
Which date would you like to start your vacation?

FIGURE 15-3.
Faking PTOBot in Slack

The learning started immediately. Even though I work in an environment where people are used to working with bots, I got a big variety of answers:

1. Not now
2. Mar 12th
3. 5/20/2017
4. Wait a min, let me confirm

Just from this small experiment I learned that getting the users to give me the answer in a way that is easy to digest might be a challenge. I will need to handle edge cases that are not actual dates. There might even be a "wait" intent that is now surfaced in the conversation, and we might plan a "waiting" story path in which the bot reminds the user to complete the PTO request intent after a while.

Wizard of Oz prototyping is a process that a lot of bot designers adopt as a quick way to explore behavior before writing a single line of code. Designers doing this process often identify issues like redundant questions, hidden entities and additional complexities they need to handle, and different ways humans answer questions. These answers might hint to us to use rich controls in some situations rather than plain text, or expect inputs we didn't think of accepting.

Another, even easier, way to explore a conversational interface is actually to have a conversation in person, or to observe a conversation in action. In this use case you can talk to a colleague at work and collect the information in person, or ask to listen to a conversation an office assistant has every day. You will notice phrases like "thank you" or "much appreciated" that you will need to filter out, and other types of conversational habits that recur in the conversation you are trying to reproduce. When you have a thoughtful and reflective conversation, you will find a lot of edge cases that will help you to design a better bot.

[KEY TAKEAWAY]

Having conversations in person, or observing real-life human conversations, is a great way to explore conversational interfaces.

Now that we are armed with the use cases and have done a little exploration, we can start scripting the conversation and later test it out in real life.

[16]

Conversation Scripting

Love without conversation is impossible.

—**MORTIMER ADLER**

In this chapter we will script the conversations our B2B and B2C bots will have. We will break down the conversations into *flows* (sometimes called *stories*), and detail the entities we want to extract. We will start to map edge cases and error handling, and will decide when to use plain text and when to use rich controls.

At a high level, here is the flow we would like to facilitate with both bots (Figure 16-1).

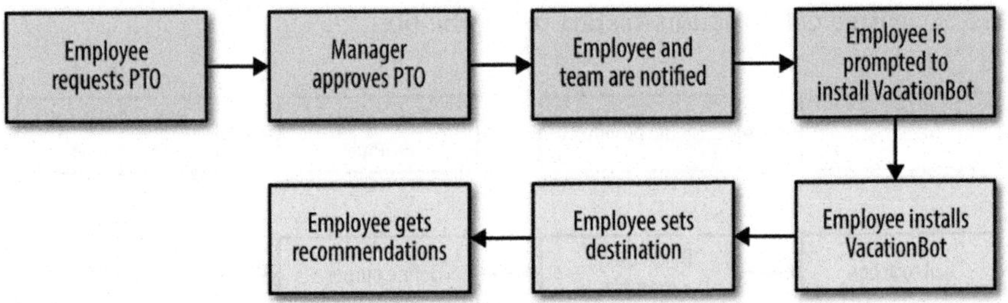

FIGURE 16-1.
A high-level view of the PTOBot/VacationBot flow

I have highlighted the PTOBot part in blue and the VacationBot part in peach. We will now outline the specific flows that compose the use case for each bot.

Outline of Flows

The first step we are going to take is exploring the different flows that compose our bots' use cases.

ONBOARDING

As previously discussed, a thoughtful onboarding could be the difference between a successful bot engagement and an abandoned bot. In this section we will try to follow the best practices of demonstrating value and providing a call to action.

> [KEY TAKEAWAY]
>
> Thoughtful onboarding could be the difference between a successful bot engagement and an abandoned bot.

PTOBot

The PTOBot onboarding is slightly complex. It starts with the person who installed the bot, and continues to the team the bot was added to (Figure 16-2). I will highlight 1:1 interactions in blue and team interactions in peach. The pattern we would like to follow is the one of a dual onboarding script. The first will end with prompting the installing user to add the bot to the relevant team channel; the second is a team onboarding script that introduces the bot to the team and ends with a call to action to start using the bot.

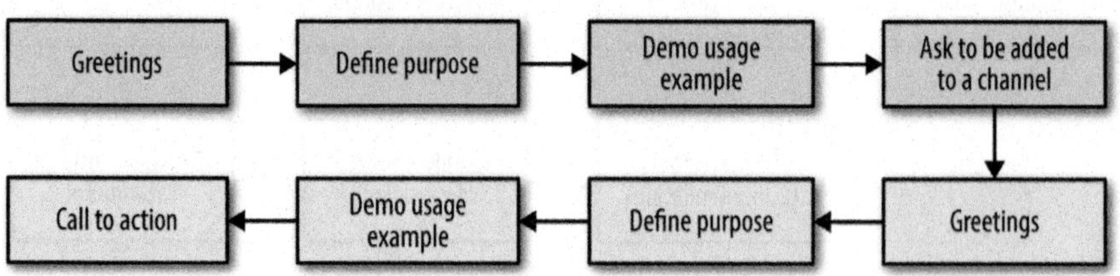

FIGURE 16-2.
PTOBot onboarding flow

VacationBot

In this flow the user will connect to the bot, and the bot will introduce itself, demo its capabilities (making the user understand its value), capture where the user is traveling to, and ask for permission to send notifications of fun activities to the user (Figure 16-3).

FIGURE 16-3.
VacationBot onboarding flow

Note that collecting the destination might trigger an error path. If the destination is not captured correctly, we should trigger an error flow (described later in this chapter) and then try again to collect the required information.

MAIN FLOW

In this flow we will describe the main happy path. This encapsulates the main functionality of the bot, without errors or divergences.

PTOBot

PTOBot's main story, a PTO request, involves a few actors: the employee, the manager, and the team (see Figure 16-4). I will use the same color notation for 1:1 and 1:many as in the last story.

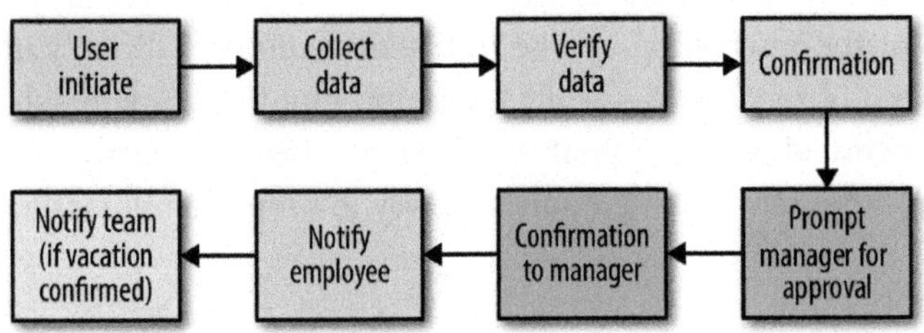

FIGURE 16-4.
PTOBot main flow

As you can see, our story starts with an employee prompting the bot to start a PTO request workflow. The bot collects the required information (we will define the required information later in this chapter), verifies that all the required data has been provided, and sends the user a confirmation of the request details. It then sends the employee's manager a request summary and prompts the manager for approval. Following manager approval/rejection, the bots notifies the user of the outcome and, if the request was approved, notifies the team.

Looking at this flow, we are facilitating a complex and usually repetitive workflow that does not require a lot of work in each step. The alternative to this use case in the traditional world would be to go through a dedicated system or fill in a paper form—so our hope is that, with this bot, this process will be a much more pleasant experience for employees.

VacationBot

VacationBot's main flow is of the in-vacation experience. It involves the user and the bot in a 1:1 interaction (Figure 16-5).

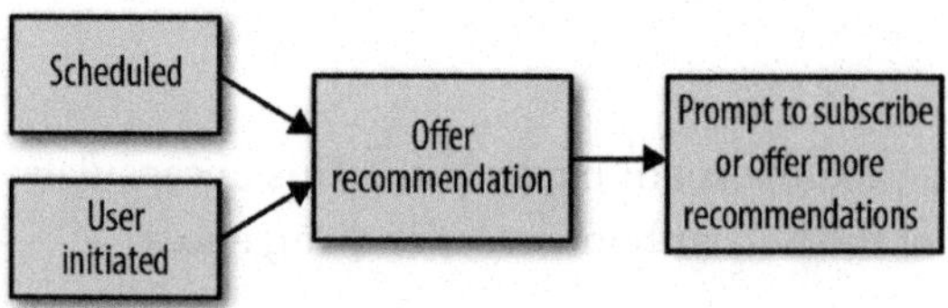

FIGURE 16-5.
VacationBot main flow

This story can start from two endpoints: the bot can provide a recommendation based on a scheduled subscription (set by the user in the onboarding) or the interaction can be initiated by the user directly in chat. At the end of the flow we would like to prompt the users to subscribe to a scheduled notification, if they have not done so already, and to offer them more recommendations as a way to keep the interaction going.

HELP

The help flow aims to support the user in case they need assistance in the main flow of the bot. This flow can kick off by the user asking for help, or the bot understanding that the user needs help due to errors in the main flow.

PTOBot

The help story for PTOBot starts with a 1:1 engagement and ends with adding a human to the conversation to resolve the issue (Figure 16-6). Because PTO-IT is a mission-critical HR system, if contextual help does not resolve the problem it's important to provide human assistance, either from tech support or HR.

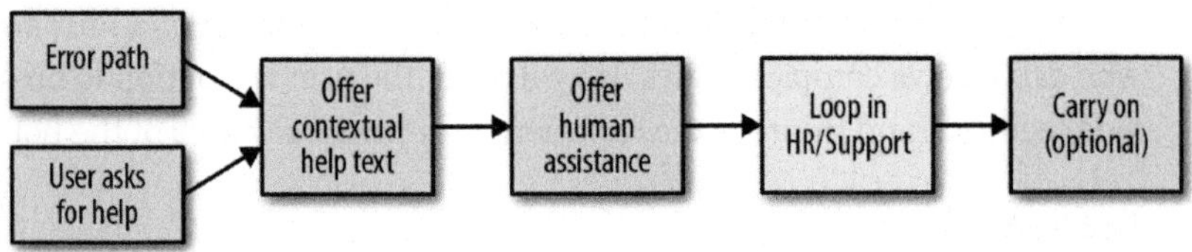

FIGURE 16-6.
PTOBot help flow

We would want to offer contextual help text in this flow. If the user gets stuck setting the start date for the PTO, for example, the help text might look like "Please enter a start date for your PTO, using the following format..."; asking for help at the beginning of the flow might just result in a repeat of the onboarding example usage script.

Note that I added an optional step of *carrying on*. A help flow can be triggered in the middle of another flow—for example, a user can ask for help when entering a description for a PTO request. You will see this pattern in several other flows in this chapter. It is up to the business logic that you outline to determine whether it makes sense to carry on with the original conversation (and pick up where you left off) or not.

VacationBot

VacationBot is less mission-critical, so we will provide only automatic measures to assist the user (Figure 16-7).

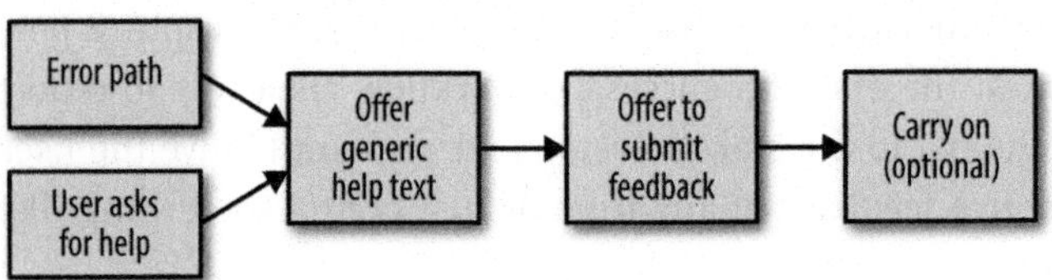

FIGURE 16-7.
VacationBot help flow

You might have noticed that we are offering generic help text. We do this because most of the steps in this bot do not require complex user inputs, so just repeating the onboarding example usage script should be enough to help the user.

Because our help path does not end with the intervention of a human who can resolve complex issues, we will offer the user the option to provide feedback and improve the experience for future releases of the bot.

FEEDBACK

As discussed previously, feedback is an important part of a conversation—it gives the user the ability to share valuable information with the bot's designer. There are several entry points to feedback (see Figure 16-8); let's describe them now.

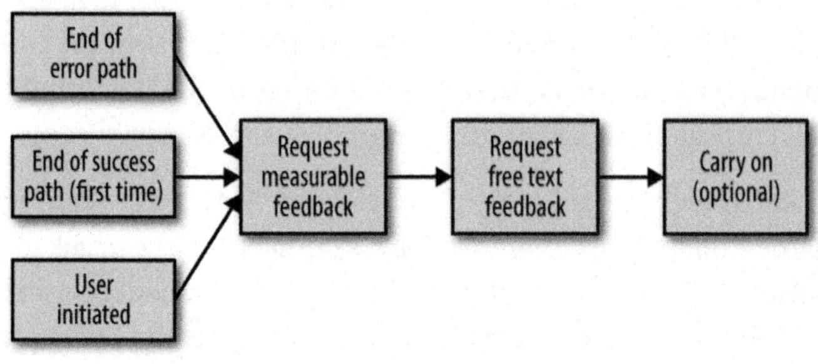

FIGURE 16-8.
Feedback flow

This flow is more or less the same in both PTOBot and VacationBot.

We want to surface feedback at several points. We want to capture negative feedback when an error flow ends (asking the user what went wrong, and whether the issue was resolved), we want to capture positive feedback at the end of a successful workflow (as a way to report the value of the bot to the stakeholders), and we want to give the user the ability to give feedback at any time by just typing something like "feedback."

Capturing positive feedback is sometimes as important as capturing negative feedback. You can use this type of feedback as client testimony, to prompt the user to provide a good review or rating in a directory listing, and to encourage users to share the bot with others. This process is used quite effectively in mobile games, many of which ask the user to rate the game after their first win. Capturing the user's happy moments can be very valuable, when done right.

[**KEY TAKEAWAY**]

Capturing positive feedback is sometimes as important as capturing negative feedback.

ERROR HANDLING

Error handling happens a lot with bots, mainly because of unexpected user inputs. I use the term *error* quite loosely—errors are all the inputs that the user enters that are not a part of the happy flow.

Examples of such inputs can include:

1. Thanking the bot
2. Cursing the bot
3. Asking the bot to do a task that the bot cannot perform
4. Random questions
5. Expressing an intent or providing an entity in a form that the bot cannot interpret with sufficient confidence (for example, "I wanna take a vacay" if the bot is not trained to map this wording to the PTO request intent)
6. Random inputs
7. Chitchat

The key with invalid inputs is to map a canned response, or several randomized responses, to a given set of inputs, and to apologize and escalate if applicable (see Figure 16-9).

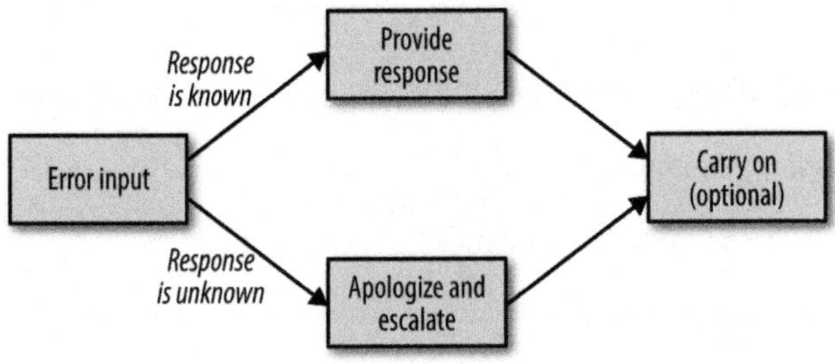

FIGURE 16-9.
Error flow

Here are a few canned response examples:

1. Thanking the bot: "Thank you!" "I appreciate it!" "I am blushing."
2. Cursing the bot: "Sorry to hear you say that," "I am sorry you feel that way," "This is not appropriate."
3. Asking the bot to do a task that the bot cannot perform: "I am sorry, I cannot complete that task."
4. Random inputs: "Sorry, I do not know the answer to that," "Not sure I got that."
5. Chitchat (e.g., "How are you doing?"): "I am fine," "Doing well," "Good, thanks."

If the input suggests that the user either is not happy or is very happy, you might want to pull in the feedback flow.

Supporting chitchat can take a lot of design effort, so avoid getting distracted into spending a lot of time on that, but supporting simple and common inputs like "How are you doing?" might provide a nice experience.

This process is very similar in both PTOBot and VacationBot, with the distinction that VacationBot does not provide escalation to a live human. If it cannot respond to the input, VacationBot can just say "Sorry I could not help, I have recorded this error in order to improve future experiences," and move on.

Intent Mapping

Here are the intents exposed by the bots and a sample of keywords that can initiate them.

For PTOBot:

- PTO request: "PTO," "Hi," "Hello," "👋," "Start," "Vacation"
- Help: "Support," "Help," "Not sure," "What?"
- Feedback: "This is great," "This sucks," "Feedback"

For VacationBot:

- Get activities: "Recommendations," "Hi," "Hello," "👋," "What's up," "News"
- Help: "Support," "Help," "Not sure," "What?"

- Feedback: "This is great," "This sucks," "Feedback"

Entity Mapping

Entities are variables we want to collect from the user. These entities can be collected using plain text or rich controls. Let's map the entities of our already outlined stories.

For PTOBot, the PTO request entities are:

- Start date {Date}
- End date {Date}
- Description {Text—up to one paragraph}
- Approved {Yes/No}

For VacationBot, the vacation activity entities are:

- Destination {Location}
- Schedule notifications {Yes/No}
- When to send the notification {Morning/Evening}

The purpose of a conversation, from the perspective of the entity-collection process, is to reach a mutual understanding about each of these entities. The bot can ask the user to input these entities or can offer them as options with rich controls. In the next section we will make that call.

Scripting Sample Bot Outputs

In this step we will create mock conversations and provide an example of each flow. For most inputs and outputs we will need to create multiple permutations so that the bot conversations will seem more natural, but for this step we will just outline single examples of inputs and outputs. This exercise will give us a better idea about what we would like to conversations to look like. This is also the place to decide whether we want to extract entities, and display information, with plain text or rich controls.

ONBOARDING

Let's start with the onboarding scripts. As we discussed previously, onboarding is a critical part of your bot experience.

PTOBot

The onboarding script for our PTOBot is split into two separate conversations: the 1:1 one with the user who is installing the bot, and a team conversation.

The conversation with the installer might look like this:

> ***PTOBot***: Thank you for adding me to the team! I am an HR bot provided to you as part of your PTO-IT solution. I can facilitate PTO requests and approvals.
>
> ***PTOBot***: Here is how I work: [show a GIF of working with the bot, one image with an employee and the other with a manager]
>
> ***PTOBot***: In order for me to work well, please add me to the relevant channels - recommended channels are #HR and #general.

The conversation with the team (once the bot has been invited to the channel) might go like this:

> ***PTOBot***: Hi everyone, I am PTOBot, your new HR bot added by [Installer name]. I can facilitate Paid Time Off (PTO) requests and approvals.
>
> ***PTOBot***: Here is how I work: [show a GIF of working with the bot, one image with an employee and the other with a manager]
>
> ***PTOBot***: You can Direct Message me at any time to start a PTO request. You can find me at @ptobot.
>
> ***PTOBot***: Another way to start a PTO request is with a slash command.
>
> ***PTOBot***: Just type: /pto-it [start-date MM/DD/YY] [end-date MM/DD/YY] [description]

There are a few things to note here. First, we assume that the installer is an HR person. They are the bot's main point of contact. We will loop this person into an escalation flow if needed in the future. The first thing we want to achieve is buy-in from this person, so they add the bot to the right channels. We could use the API to automatically add the bot to several channels, but that would probably be perceived as spammy and intrusive. Having the bot added by an HR person is a good signal to the team that this is the recommended way to request PTO. You might have also noticed that we use the installer's name in the team script, reinforcing this link.

When the bot is added to a channel (it receives an event each time this occurs), it introduces itself to the team and educates them about how to use it. You might also have noticed the decision to post a GIF instead of describing how the bot works—we will need to test this experience with real users, but the assumption here is that the GIF will be an easy way to demonstrate how to work with the bot.

The last thing to note is that we offer a shorthand, command line–like way to ask for PTO. We anticipate that tech-savvy users might find it more efficient to type in a single line to kick off a PTO request, like this:

> *User*: /pto-it 04/21/17 04/31/17 Holiday vacation

VacationBot

Now let's turn to our VacationBot. The onboarding of VacationBot requires us to collect the trip destination entity, and also to end with a call to subscribe to notifications about activities. The underlying assumption here is that this bot is installed straight after the PTO request is approved. PTOBot will suggest that users install VacationBot to learn about fun activities at their destination after their PTO has been approved.

Here's our example onboarding conversation script (between VacationBot and the user):

> ***VacationBot***: Thank you for connecting!
>
> ***VacationBot***: I am your friendly VacationBot. I can provide you with news about activities happening while you are on vacation. I will even include your corporate discounts, provided to you by PTO-IT!
>
> ***VacationBot***: Our records show that you are heading out on 04/21/17.
>
> ***VacationBot***: Please provide your destination so we can get started.
>
> *User*: Cancun Mexico
>
> ***VacationBot***: Fantastic place for a vacation! Here are a few cool attractions in the area:
>
> ***VacationBot***: [Carousel with 3 activities]
>
> ***VacationBot***: I can send you fun new activities every day. Would you like that?

VacationBot: [Yes, every morning] [Yes, every evening] [No thanks] *(Quick Replies)*

{ Yes, every evening }

VacationBot: Fantastic, will do. Have a great vacation! You can always say "Recommendations" to get more recommendations at any time.

{No}

VacationBot: That is fine, you can always say "Recommendations" to get more recommendations at any time. Have a great vacation!

VacationBot: [Recommendations] *(Quick Reply)*

Again, there are a few things to note in this example. The first is that we immediately state the name of the bot, its purpose, and also its association with the PTO-IT offering. We do this as preparation for what comes next. In the next section we tell the user that we already know when they are starting their vacation. The association made before that is to make sure the user knows that information has not been leaked or acquired inappropriately. We want to provide the user with the feeling that this is all part of a service provided by the PTO-IT offering.

Next, you will notice that we are prompting the user to add their destination. This is information that is not collected by PTOBot (it's not relevant for PTOBot's operation, and it might be inappropriate for that bot to request it), so we need to collect this entity now. There is a hidden flow that might surface if the bot does not understand the destination, but for now we will assume a valid user input.

The last thing to note here is that we have started to make decisions on how to interact with the users. We have annotated the script with things like *(Quick Reply)* to mark our decision to use this type of rich interaction in the script.

This script uses a combination of Quick Replies and free text input. We use free text for open-ended questions—for example, we currently do not have a rich control that lets the user pick a location (though this might be provided by Facebook Messenger in the future), so we default to just asking the user to input their destination as free text. In other places, where there are simple choices to be made (like whether and when to get the notifications), we use Quick Replies to indicate to the user what the valid choices are. We will still need to manage the use case where the user inputs free text instead of clicking on a Quick Reply, but the assumption is that users will click on the replies most of the time, rather than typing.

> **[KEY TAKEAWAY]**
> Use buttons and Quick Replies for non-open-ended questions to indicate to the user what the valid choices are.

This conversation ends with a call to action, by showing the [Recommendations] Quick Reply. This hints to the user that their next step is to get more recommendations; this keeps the conversation going and provides users with a feeling of not being stuck or lost.

MAIN FLOW

Now we can script the main flow for each bot.

PTOBot

PTOBot's main flow is actually made up of four scripts: PTO request, PTO approval, PTO approval notification, and team notification.

Here's our example PTO request conversation script (between the employee and PTOBot in a direct message):

> *User*: Hi @PTOBot I would like to take a PTO
>
> ***PTOBot***: Happy to assist! Which date would you like to start your PTO?
>
> *User*: 04/21
>
> ***PTOBot***: Thanks, which date would you like to end your PTO?
>
> *User*: 04/31
>
> ***PTOBot***: Please provide a description for your PTO request.
>
> *User*: Holiday vacation
>
> ***PTOBot***: Great! To confirm, you would like to take PTO between the 04/21/2017 and 04/31/2017 for a Holiday vacation.
>
> ***PTOBot***: [Make Request] [Edit] *(buttons)*
>
> {"Make Request" button clicked}

PTOBot: Thank you! I will now send the request to your manager, April Underwood, to approve your PTO.

User: Thanks!

PTOBot: You are welcome.

Another version of the same script would involve the shorthand version with the slash command:

User: /pto-it 04/21/17 04/31/17 Holiday vacation

PTOBot: To confirm, you would like to take PTO between the 04/21/2017 and 04/31/2017 for a Holiday vacation.

[Make Request] [Edit] *(buttons)*

{"Make Request" button clicked}

PTOBot: I will now send the request to your manager, April Underwood, to approve your PTO.

As you can see, in both versions of this script we are using a button to confirm the request and free text to set the other entities. Because the intent of the slash command is to make the interaction short and to the point, I have removed the "niceties" ("thank you," "great," and such) from the script.

You might also have noticed that the user in the first script thanked the bot. Here we see a small snippet of the error handling use case, where the input is not critical for the workflow, but an answer like "You are welcome" is appropriate.

Now let's move on to the second part of the PTO flow.

Here's our example manager PTO approval script (between the manager and PTOBot in a direct message):

PTOBot: Hi April, Chris Messina would like to take PTO between **04/21/2017** and **04/31/2017**. If approved, Chris will have **5 days in surplus** remaining.

PTOBot: [Approve] [Reject] *(buttons)*

{"Approve" button clicked}

PTOBot: Thank you, I will notify Chris Messina that the PTO request has been approved.

Note that we are pulling in information that was not supplied by the user. PTOBot is part of the PTO-IT solution, and the assumption is that the bot has access to extra information, like the employee's PTO balance, that can help the manager make a decision on whether to approve the request or not.

Next, let's go back to the employee to confirm the manager's approval.

Our PTO approval notification script (between the employee and PTOBot in a direct message) looks like this:

> ***PTOBot***: Good news! Your manager, April Underwood, has approved your PTO request: *(formatted in a message attachment)*
>
> ***PTOBot***: Would you like me to send a notification in the #PTO channel?
>
> ***PTOBot***: [Notify Team] [Skip] *(buttons)*
>
> {Any button clicked}
>
> ***PTOBot***: If this PTO is a vacation, you might be interested in our new VacationBot offer. This is a Facebook Messenger bot that can inform you about activities and corporate discounts at your destination.
>
> ***PTOBot***: Here is the link to install it - *[link to vacation bot]*

As you can see, we are using a formatted message attachment for this notification. The aim is to provide the employee with a receipt-like experience that will hopefully increase their trust in the system.

We also offer the employee a way to notify the team of their upcoming PTO. We will script that part in the next section. Lastly, we suggest that the employee install the Facebook Messenger–based VacationBot. This will be the segue to convert PTOBot users into VacationBot users. We are using a link to VacationBot; this link will be provided by Facebook Messenger when we publish this bot on their platform. Clicking on the link will take the user straight into a conversation with that bot.

The last part of PTOBot's main flow is the notification to the team.

The PTO notification script (PTOBot posting in the #pto channel) looks like this:

PTOBot: PTO alert:

PTOBot: Employee: Chris Messina

PTOBot: When: 04/21/17 04/31/17

PTOBot: Approved by April Underwood

Note that we have omitted the description from the notification—as this is a public channel, we do not want to share that potentially sensitive information.

This ends the main script for our PTOBot. You'll notice that we demonstrated team engagement as well as personal one-on-one engagement with the bot. This flexibility is common in work environments and is useful for facilitating complex business processes.

Now let's move to our VacationBot and its main flow.

VacationBot

Our VacationBot has a simpler main flow. It is driven by a scheduled task set by the user in the onboarding script, or at any time by the user clicking on the "Recommendations" Quick Reply. Here is how this script goes:

> *VacationBot*: Hello again! Here are a few new cool attractions in the area:
>
> *VacationBot*: [Carousel with 3 activities]
>
> *VacationBot*: [More] [Schedule notifications *(if not subscribed)*] *(Quick Reply)*

This is very similar to the onboarding script. Notice that we are providing the user with a way to subscribe to get notifications, in case they did not do so initially in the onboarding phase. Like with many consumer services, a lot of the magic here is around the simplicity of the flow and the content displayed in the carousel. We will experiment with the design of the carousel itself in the next chapter.

In real life, there might be a lot of additional complexities in the main scripts of both bots. For example, our VacationBot might have activities divided by categories, and our PTOBot could have a path where the user is getting into a negative PTO balance. We have simplified the use cases of both bots not only in order to make this book slim enough for you to carry around, but also because the design methods we've used here can solve these more complex use cases by repeating the steps we have taken here.

HELP

Now let's turn our attention to the scripts for the help flow.

PTOBot

For our PTOBot, the help script for the vacation approval process has three parts to it: generic, context-specific, and human intervention.

Here's our generic help script (PTOBot and employee in a direct message):

> ***PTOBot***: Here is how I work: [show two GIFs of working with the bot, one with an employee and the other with a manager]
>
> ***PTOBot***: You can direct message me at any time to start a PTO request. You can find me at @ptobot.
>
> ***PTOBot***: Another way to start a PTO request is with a slash command.
>
> ***PTOBot***: Just type: /pto-it [start date in MM/DD/YY format] [end date in MM/DD/YY format] [description]
>
> ***PTOBot***: [Need human help] *(button)*

Notice that this is very similar to the onboarding script provided by the bot. The target here is to provide generic assistance in case the user has forgotten how to work with the bot. This script will usually appear if the user types "help" before starting a PTO request.

We have added a button at the end of the script that the user can click to seek support from a human, in case this generic help script does not suffice. We will script the human intervention script shortly.

Next, here's an example of a context-specific help script (PTOBot and employee in a direct message):

> ***PTOBot***: You are now entering a start date for your PTO request. Please pick the first business day of your PTO and use the following date format: MM/DD/YY.
>
> ***PTOBot***: [Need human help] *(button)*

This is an example of a help script that might be produced when the user types "help" while creating a PTO request if they run into problems entering their PTO start date. You can create a set of contextual help snippets and supply them at the right time when a user needs help. If these help snippets do not work, we default back to requesting human assistance, like in the case of the last script. Let's explore that script now.

Our human assistance script might look like the following (between PTOBot, the employee, and HR support in a multi-party direct message):

> {"Need human help" button clicked}
>
> ***PTOBot***: Hello HR support. Chris Messina is running into issues creating a PTO request. I was not able to resolve this issue, so I am connecting us all so it can be successfully resolved.
>
> *Peter Skomoroch*: Hi Chris! My name is Peter and I am a support engineer with your People Ops department. How can I help?
>
> *Chris*: Hi Peter, I have a question regarding...

Looping a human into a conversation can be done in two ways:

Bot as a router
: The bot is backed up by a human who takes over and talks to the user. The user thinks they are still talking to software but in fact they are talking to a human.

Bot as a connector
: The bot connects the user with a human to resolve the issue.

We have chosen the bot as connector method because we want to create a "personal touch" experience for the human-tier support. Looping a real person into the loop will hopefully promote the impression of a high-end brand together with a strong feeling of service quality provided by the internal HR support staff.

At the end of the human interaction flow, the user can be redirected back to the conversation or post feedback on the system:

> ***PTOBot***: I hope the issue was resolved successfully. What would you like to do next?
>
> ***PTOBot***: [Continue with PTO request] [Feedback] *(buttons)*

VacationBot

The help flow for our VacationBot is composed of two scripts, one generic for any user inputs, and the other specific for problems capturing the destination entity.

The generic help script (between VacationBot and user) would be something like this, following the user typing "help":

> ***VacationBot***: Happy to help! I can give you recommendations of activities in Cancun. Just say "recommendations" at any time and I will post new and exciting vacation activities. For additional support please email support@pto-it.com.

And here is what the specific "destination entity capturing" help script might look like (between VacationBot and user):

> ***VacationBot***: Please provide a city and state in the USA, or a city and country worldwide. For example, "Los Angeles, California" or "Cancun, Mexico". For additional support please email support@pto-it.com.

Because our VacationBot is not considered mission-critical, we will not provide a live human support escalation path. The bot interaction is also very simple, and we do not anticipate a lot of issues that will need support. We do provide an email address for general support as part of the service.

FEEDBACK

The feedback scripts for both bots can be more or less the same (between PTOBot/VacationBot and user):

> ***Bot***: We would love to get your feedback!
>
> ***Bot***: How would you rank your experience:
>
> ***Bot***: [Great] [Good] [Poor] [Terrible] *(buttons)*
>
> {Any button clicked}
>
> ***Bot***: Please provide your verbal feedback in up to one paragraph.
>
> *User*: Love this bot, I really want to take it to every vacation!
>
> ***Bot***: Thank you, your feedback was submitted.

In this script you'll notice a combination of rich interaction and free text inputs. We want to capture measurable feedback on the experience, as well as encourage the user to share additional information in an open way.

ERROR HANDLING

We will divide error handling into four categories:

Chitchat
: Acceptable inputs that are not relevant to the conversation but are nice to acknowledge

Entity extraction issues
: Invalid inputs while trying to capture a needed input from the user

Abuse
: Unacceptable user inputs that should be stopped

Generic
: A set of error messages that are the default when none of the others is a good fit

Here are a few chitchat script examples:

> *User*: Thank you!
>
> ***Bot***: You are welcome.
>
> *User*: How are you doing?
>
> ***Bot***: I am doing well. Thanks for asking.
>
> *User*: You are great!/ You suck!
>
> ***Bot***: Would you like to leave feedback?
>
> ***Bot***: [Leave feedback] [Skip] ***(buttons)***

There are many possible examples of these micro-conversations. In some bots the chitchat component is very important, because it creates a strong brand attachment. As Andy Mauro, the CEO of Automat.ai, shared:

We built an influencer marketing bot called KalaniBot. The normal way that influencer marketing is measured is by engagement (views, likes, comments), with comments being the most engaged and valuable. Keep in mind that the average fan sends KalaniBot 14 messages on average and so is spending multiple minutes interacting relative to the short amount of time it takes to view and like a post. Conversations are better than short-lived campaigns—an Instagram post has a short 24–48-hour lifespan, whereas the InfluencerBot is still going strong a month later. Here are examples of comments we saw from users in conversation with KalaniBot:

—I love u and I think ur bot is amazing

—I love this bot

—Omg your bot is amazing girl. Too bad i cant speak to the real you

—It's AWESOME!!

—This is amazing

—This is so cute u go girl

—This is kewllll

Most users just want to be acknowledged, and it is very important that the bot does that as part of its functionality.

Trying to guess every type of chitchat the user might make is a never-ending task. You might want to limit the amount of chitchat you support, especially in task-led conversations, where it can get in the way of getting things done. Avoid dead ends, but do not spend all your time trying to think of every possible input. At a certain point you will need to default the user to the generic error messages.

[KEY TAKEAWAY]

Avoid dead ends, but do not spend all your time trying to think of every possible input. At a certain point you will need to default the user to the generic error messages.

Moving on, let's look at a very basic entity extraction error script example:

> ***Bot***: Sorry, I did not understand that.
>
> ***Bot***: <display relevant contextual help>

This is an example of one flow pulling in another flow. We are in an entity extraction flow (let's say, in the main use case), and the user inputs invalid text. The bot then provides a simple apology and error message, and pulls in the help script relevant to this specific context.

The result in real life might look like this:

> ***VacationBot***: Please provide your destination so we can get started.
>
> *User*: On the moon!
>
> ***VacationBot***: Sorry, I did not understand that.
>
> ***VacationBot***: Please provide a city and state in the USA, or a city and country worldwide. For example, "Los Angeles, California" or "Cancun, Mexico". For additional support please email support@pto-it.com.

Notice the composition of the scripts—we start with an *onboarding* script, get invalid user input, then move to an *entity extraction issues* script, and end up with a *destination entity capturing* script. You can start to see how all of our scripts work together to provide a comprehensive conversation experience.

Now, you might think it is strange that we single out abuse as something we will need to deal with, and it is true that, in our use cases, it is very unlikely that our bots will suffer major abuse. But a lot of bot builders do report abusive inputs, from hate speech to obscenities. If your bot is a brand bot, or is representing a famous person or a service, there is a high likelihood that you will need this type of script.

Here is a real-life example of a bot planning for abusive use cases, told by Greg Leuch, head of product at Poncho:

> We launched with a limited timeframe, limited resources, and a lot of pressure to succeed as a showcase example of the Messenger platform. The negative feedback for Microsoft's Tay chatbot was hanging over us as we prepared to launch. We spent a lot of time building

anti-trolling tools (naughty word filters), anti-abuse tools (allowing Poncho to ignore abusive users), and conversations for a variety of the anticipated trolling, testing, & abuse.

Coming back to our scripts, I decided to default to a set of scripts that will hopefully stop the abuse as soon as it starts. For example:

User: <Curse>

Bot: Sorry you feel that way.

<Consider pulling in a feedback script>

User: <Obscenity>

Bot: Sorry, I am not into these types of conversations.

I indicated that we might want to consider pulling in a feedback script after handling a curse (or negative comment) from the user—this is a good idea if we think the curse is actually a result of frustration from a failure to achieve a task.

I also talked to the Kik team about abuse on their platform. They told me they recommend a "three strikes and you're out" rule, in which the bot bans the user after three abuse warnings. This can be a good model for bots on Kik that interact with teens on a regular basis.

A generic error is a last resort we want to fall back on, when all the rest of the error strategies have failed or were not a good match to a situation. The process here is to provide a few versions of apologies, and to move to either feedback or help. For example:

Bot: Sorry, I can't help you with that.

Bot: [Feedback] [Help] *(buttons/Quick Replies)*

Bot: Apologies, I don't know what to do with what you told me.

Bot: [Feedback] [Help] *(buttons/Quick Replies)*

Bot: Hmm, not sure what to do next.

Bot: [Feedback] [Help] *(buttons/Quick Replies)*

These scripts will be similar in both PTOBot and VacationBot. The implementation of feedback or help in the two bots might be different, but that is the responsibility of the specific feedback or help scripts for each bot.

Remember that all of these scripts might be very different for the bot you are trying to design, not only in functionality, but also in tone and personality. Also remember that these are simply sample scripts, and that in real life you will need to provide a lot more variations, both in inputs and in outputs of the bot.

Doing scripting like this is just like creating mockups of a mobile or web application. We need to validate the scripts we have written in real life. We will also need to test out some of the assumptions we have made regarding rich interactions and see if users use them the way we thought they would.

[KEY TAKEAWAY]

Scripting is just like creating mockups of a mobile or web application. You need to validate the scripts you have written in real life.

Another thing to realize is that scripts should be evolving on an ongoing basis—we will continually monitor the success rate of our bots at achieving the tasks they were assigned to do, explore the error and help cases, read through the feedback the bots collect, and continue to optimize our bots' scripts.

Advanced bot builders hold these scripts in a content management system (CMS) and have scriptwriters go over and optimize the conversations on a daily basis. For now, let's keep the scripts as they are, and move forward to designing them and putting them in the hands of our potential users.

[17]

Designing and Testing

I love taking an idea to a prototype, and then to a product that millions of people use.

—SUSAN WOJCICKI

THERE ARE A LOT of ways to design a conversation. One option is to just expand on the Wizard of Oz technique demonstrated in Chapter 15, and mimic the scripts by impersonating the bot. While this is a very easy and quick method to get your product in front of your users and other stakeholders, it provides low fidelity when it comes to rendering rich interactions. This is because the chat platforms limit the types of rich controls available to humans. Users can post simple images, GIFs, and even videos, but they cannot display buttons, for example.

When it comes to software solutions, there are also a lot of design tools that provide you with different levels of fidelity and ease of use. There are many good options, and you should pick the ones that suit you. I have chosen two tools as examples, one for designing bots for Facebook Messenger and the other for Slack.

In the next few sections, we will go over the scripts we created in the previous chapter and use these design tools to visualize these scripts. For each script, we will try to fine-tune the wording, formatting, and other aspects of the conversation. This is an iterative process that will demonstrate design in real life.

Designing VacationBot for Facebook Messenger with Botsociety

Let's start with a tool called Botsociety (*https://botsociety.io*). Botsociety is a super-easy and quite full-featured design tool for bots. At the time of writing Botsociety supports only Facebook Messenger, but the team have told me they are planning to launch support for other platforms very soon.

After registering, we will choose a name for our bot, select a platform, and start sketching (Figure 17-1).

FIGURE 17-1.
Creating a mockup with Botsociety

Next, we'll go into the main designing area. As you can see, the tool uses a super-simple "Bot Says"/"User Says" paradigm (Figure 17-2).

FIGURE 17-2.
Drafting the conversation script ("Bot Says"/"User Says")

Clicking on the "Bot Says" button provides you with a choice of the common types of output bots on the Facebook Messenger platform can provide (Figure 17-3).

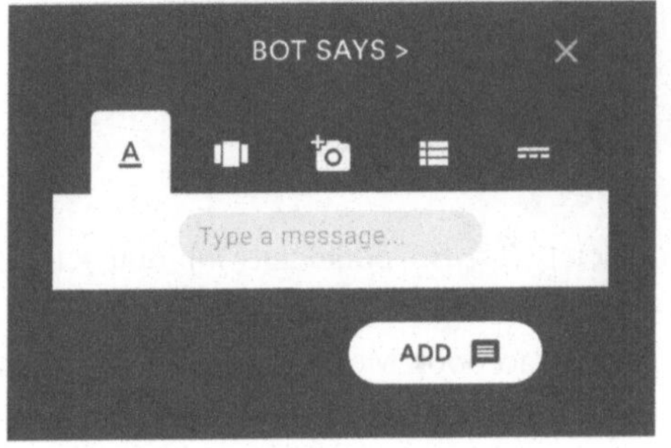

FIGURE 17-3.
Output types available to bots on Messenger

Similarly, clicking on "User Says" offers you a choice of the types of input available to users on Facebook Messenger (Figure 17-4).

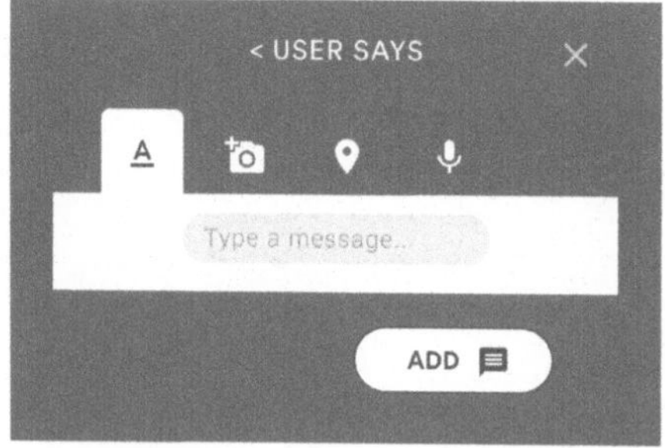
< USER SAYS
Type a message...
ADD

FIGURE 17-4. Input types available to users on Messenger

From a quick browse of the elements on both sides of the conversation, it looks like we have everything we need to start designing our scripts. So let's give it a try! I have taken the onboarding script outlined in the previous chapter for our VacationBot and entered the first part of it into the tool. Immediately, I notice an issue with the original script—it's so long that it runs below the fold (meaning you need to scroll read it all). This makes for a bad user experience, as it is hard to understand at a glance what the bot is and what it wants from the user (Figure 17-5).

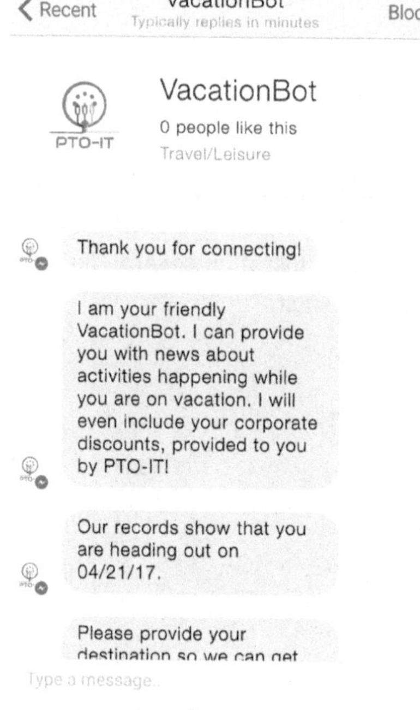
JL
2:46 PM
89%
Recent
VacationBot
Typically replies in minutes
Block
VacationBot
0 people like this
Travel/Leisure
PTO-IT
Thank you for connecting!
I am your friendly VacationBot. I can provide you with news about activities happening while you are on vacation. I will even include your corporate discounts, provided to you by PTO-IT!
Our records show that you are heading out on 04/21/17.
Please provide your
Type a message...

FIGURE 17-5. VacationBot's onboarding script—the user needs to scroll to read all of the text

I have a bad feeling about this: if users do not see the entire value proposition up front, they might just back out of the conversation. Let's make it shorter (Figure 17-6).

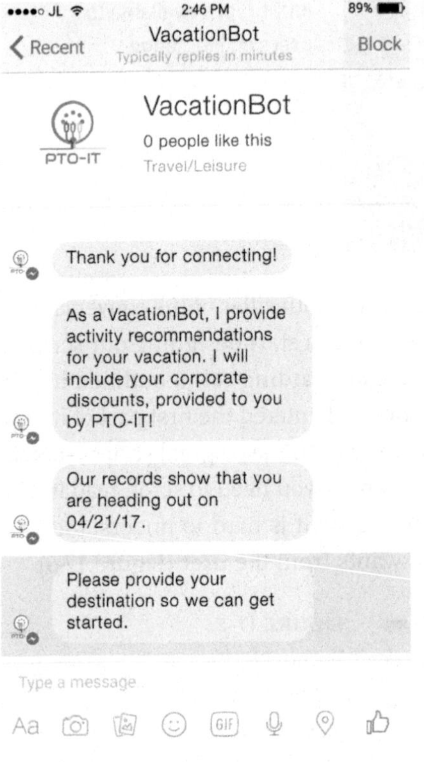

●●●●○ JL
2:46 PM
89%
Recent
VacationBot
Typically replies in minutes
Block
PTO-IT
VacationBot
0 people like this
Travel/Leisure
Thank you for connecting!
As a VacationBot, I provide activity recommendations for your vacation. I will include your corporate discounts, provided to you by PTO-IT!
Our records show that you are heading out on 04/21/17.
Please provide your destination so we can get started.
Type a message
Aa
GIF

FIGURE 17-6.
A shortened version of the onboarding script

I made the bot's value proposition a little more concise (which is a good thing on its own), and now everything fits well within the window without scrolling. We will, of course, need to test it on several devices, both mobile and desktop, but this is a good start.

[KEY TAKEAWAY]

The difference between a good experience and a poorly executed one can be in the small details, such as how long the text is and whether the user has to scroll to read all of it.

Some designers will prefer to do all of their scripting in tools like this, for the benefit of seeing immediately how the script looks in real life—if you feel more comfortable doing so, please do. I prefer to start with written scripts, as it enables me to really think about the flow in the context of multiple scripts and use cases. It is also easier to cut, paste, and share initial scripts written as text rather than as a GIF, which is the output of this tool.

Now we can continue with the script and see if the rich interaction we envisioned works well. Botsociety provides you with the ability to render more than plain text: it also enables you to render rich controls like carousels. Near the end of our onboarding script, we have a section where we demonstrate the value of the bot with a carousel of activities at the vacation destination. We might need to wait for user feedback, but I think the outcome is quite nice (Figure 17-7).

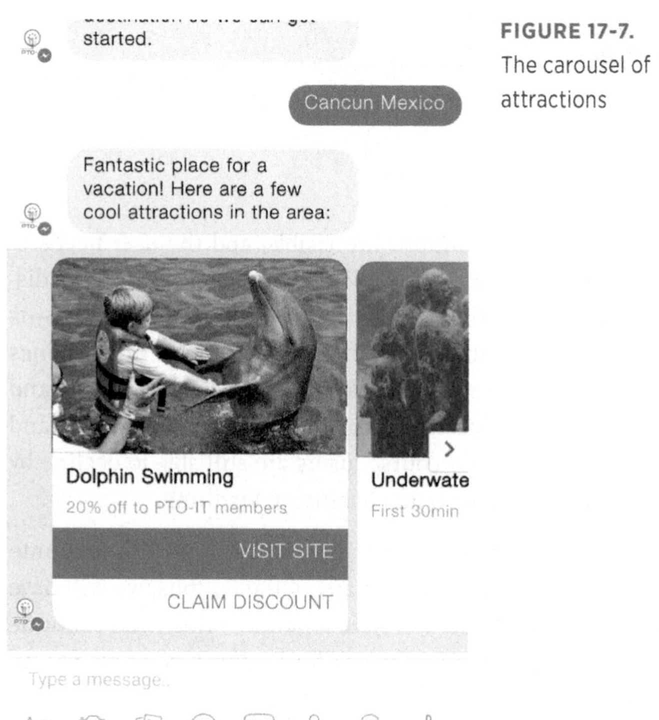

FIGURE 17-7.
The carousel of attractions

Finally, we can explore the call to action at the end of our onboarding script (Figure 17-8).

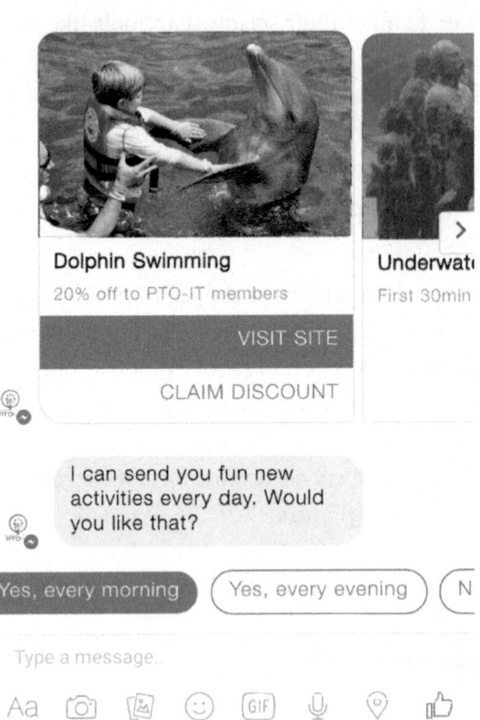

FIGURE 17-8.
The call to action

As you can see, not all the buttons are visible, and the user needs to scroll to see all the options. This could actually be a blessing in disguise, as the last option—the one the users need to scroll to pick—is the one that we do not want them to click on (the option that declines the offer). The positioning of the UX elements, both on the screen and off, subtly encourages the user to pick one of the "right" choices and subscribe to the bot's feed. Of course, users are still able to decline by clicking on "No thanks" or just not clicking on anything.

This brings to light another consideration we will need to take into account—if the user does not click on any of the buttons, we will have to treat it as a "No thanks" after a certain amount of time, and continue on to provide the user with a way to ask for recommendations manually at any time.

By now we have tested all of the rich controls we have planned for our VacationBot, for this use case. Let's finish up the onboarding design, and pick the "No thanks" Quick Reply (Figure 17-9).

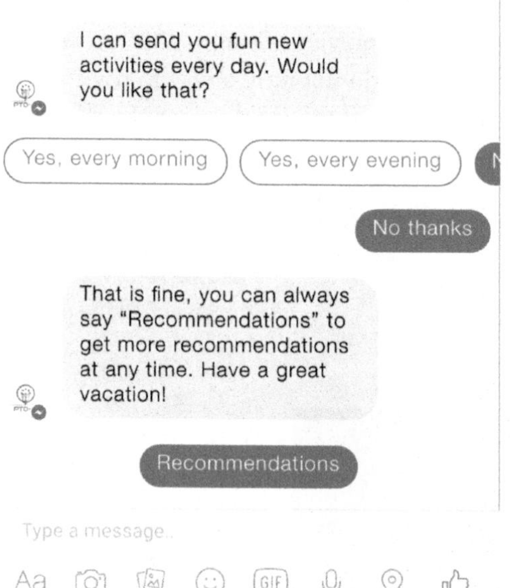

FIGURE 17-9. Handling the "No thanks" reply

It looks like we can do a little better—we are missing the opportunity to let the user schedule reminders at a later stage. Let's add that now (Figure 17-10).

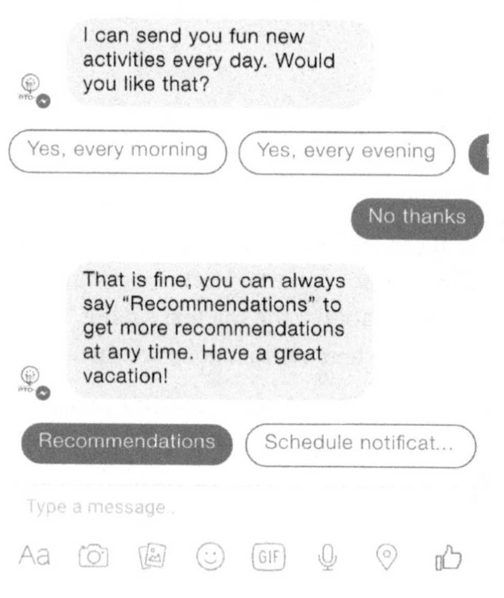

FIGURE 17-10. Adding a Quick Reply to schedule notifications

This looks much better now, and the user has another option to subscribe to the service. Note that in the previous design, after users have declined the offer to subscribe, they do not have a way to back out of that decision. It is always recommended to give the user the option to reconsider and do the right thing.

Now we will implement VacationBot's main flow (Figure 17-11).

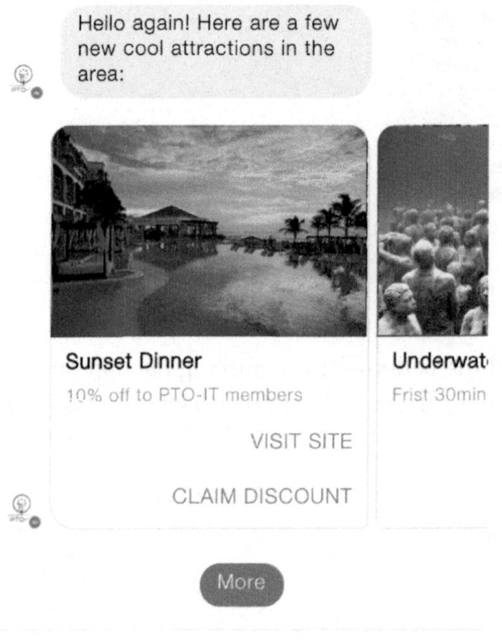

FIGURE 17-11.
Implementing the main flow

I think it looks pretty good. Remember, we will have a "Schedule notifications" option in case the user has not done so. Here, we assume the user has subscribed to receive notifications; the top text saying "Hello again!" would not be visible in cases where notifications are turned off.

Now let's design the help script. We will add Quick Replies at the end of the help text to prevent a dead end (Figure 17-12).

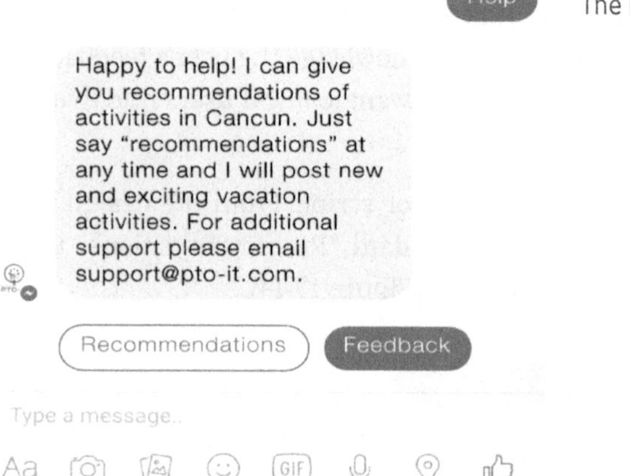

FIGURE 17-12.
The help script

This is better than the original script because it is more consistent and always provides the user with options for what to do next.

Now let's implement the feedback script of our VacationBot (Figure 17-13).

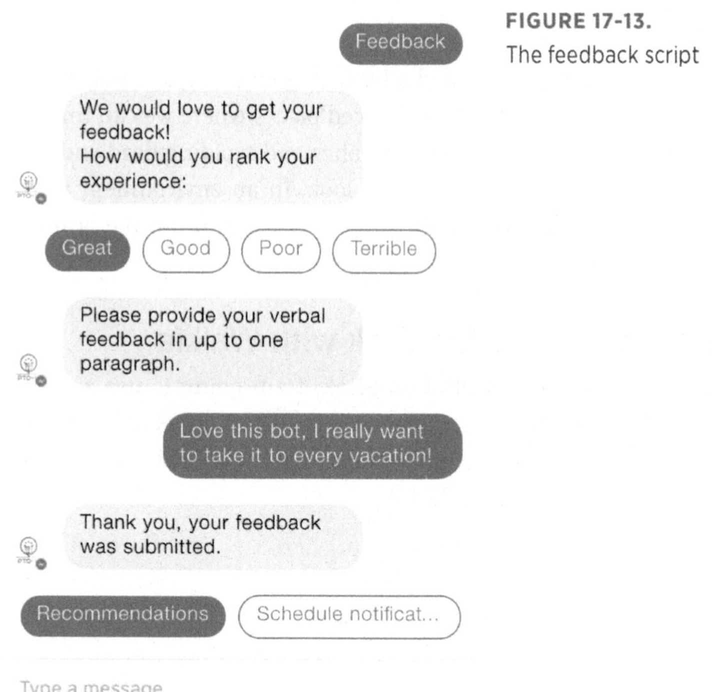

Feedback
We would love to get your feedback!
How would you rank your experience:
Great
Good
Poor
Terrible
Please provide your verbal feedback in up to one paragraph.
Love this bot, I really want to take it to every vacation!
Thank you, your feedback was submitted.
Recommendations
Schedule notificat...
Type a message...

FIGURE 17-13.
The feedback script

I have added the consistent ending that gives the user a hint about what to do next. I really like the outline of the Great...Terrible Quick Replies. One issue is that the bot does not acknowledge the user's feedback rating; this might be fine, but we might want to test if users find that awkward or not.

Finally, we will design a generic error script. From the design up to now, I already know to add the standard "Recommendations" Quick Reply at the end of the conversation (Figure 17-14).

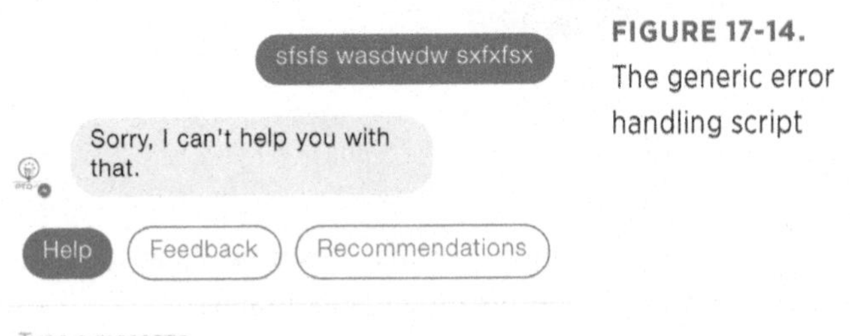

FIGURE 17-14. The generic error handling script

The last thing I've noticed is that the logo looks really bad—the text is not visible, and it looks small. We will fix this for both bots later in this chapter.

As you can see, we have learned a lot from just doing a simple visualization of our scripting. We have noticed places where we can improve user engagement by adding Quick Replies and avoiding dead ends. We have also seen how the rich controls look, in an environment that is close to real life, and have modified our text to improve the layout of our conversations.

Designing PTOBot for Slack with Walkie

In order to design our PTOBot on Slack, I am going to use a design tool called Walkie (*https://walkiebot.co*). Walkie is a flexible and a feature-rich tool that lets you script multiple flows. It all starts with setting up your bot and user (Figure 17-15).

FIGURE 17-15.
Getting started with Walkie

After saving the settings, we go into a Slack-like user experience (Figure 17-16).

FIGURE 17-16.
Walkie's Slack-like UI

On the left there is a list of bots that we've created (in this case, PTOBot), then there is a list of flows which are distinct scripts (PTO approval, for example). The main area to the right is the design section, with a place to enter user and bot inputs. Clicking on the "User" button toggles between bot and user. There is also a control at the bottom right to create rich interactions through message attachments. Clicking on it opens up a fully configurable message attachment, including buttons (technically called *attachment actions*; see Figure 17-17).

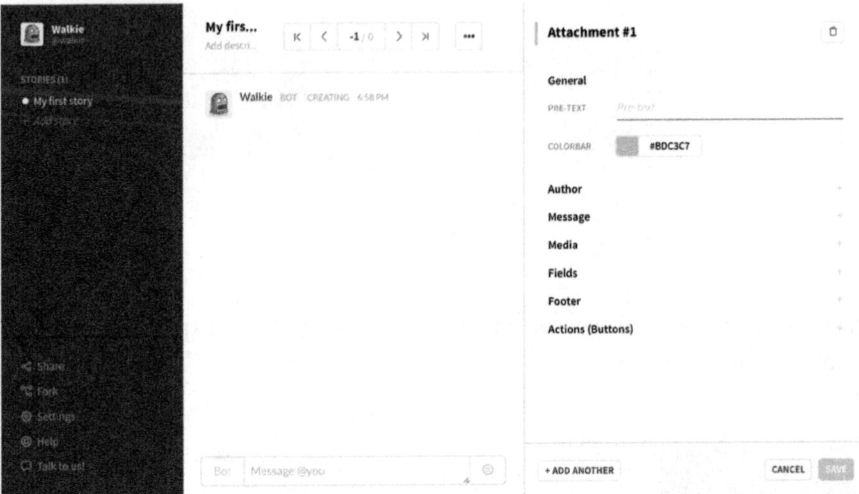

FIGURE 17-17.
Adding a message attachment

The tool does a good job of supporting multiple bots, but does not support multiple users (user personas) interacting with a bot. So, I will create a few bot configurations to work around that limitation.

Let's start with the onboarding script. The first conversation is with the user who installed the bot (Figure 17-18).

As you might recall, the onboarding script shows a GIF of the PTO request process, in order to demo the bot's usage. Showing this in a printed book is a challenge on its own, so I have cheated a little and used a screenshot of the PTO request process.

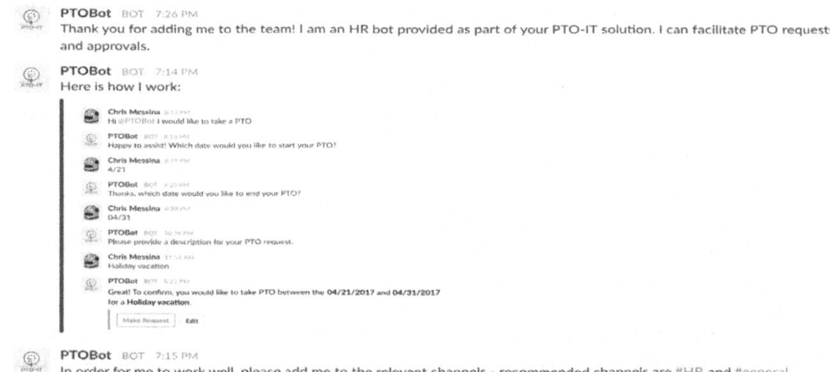

PTOBot BOT 7:26 PM
Thank you for adding me to the team! I am an HR bot provided as part of your PTO-IT solution. I can facilitate PTO requests and approvals.
PTOBot BOT 7:14 PM
Here is how I work:
PTOBot BOT 7:15 PM
In order for me to work well, please add me to the relevant channels - recommended channels are #HR and #general.

FIGURE 17-18.
The onboarding conversation with the bot installer

The script looks okay, but it suffers from the opposite problem our VacationBot had in its onboarding script—the script here seems too dry and too short, and I am not sure it is clear and actionable enough.

Let's try to fix that (Figure 17-19).

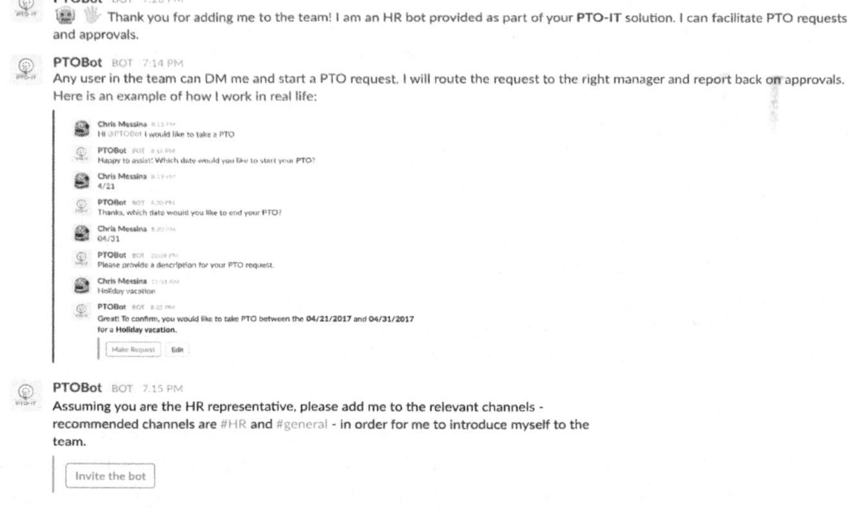

PTOBot BOT 7:26 PM
Thank you for adding me to the team! I am an HR bot provided as part of your PTO-IT solution. I can facilitate PTO requests and approvals.
PTOBot BOT 7:14 PM
Any user in the team can DM me and start a PTO request. I will route the request to the right manager and report back on approvals.
Here is an example of how I work in real life:
PTOBot BOT 7:15 PM
Assuming you are the HR representative, please add me to the relevant channels - recommended channels are #HR and #general - in order for me to introduce myself to the team.
Invite the bot

FIGURE 17-19.
Fleshing out the installer onboarding script

I added a few emojis, made the text a little more descriptive, and added a button at the bottom of the script that lets the installer invite the bot to the right channels with a single click. Because the Slack API lets us to add a bot to a channel programmatically, we can use this button to shortcut the need for the user to go into the relevant channels and invite the bot manually. We will use the API to add the bot automatically, while still giving the control to the installer, by only adding the bot after the button has been clicked.

I also started to add a color convention: blue will be *informative* (like the blue color next to the demo GIF) and green *actionable*, for actions we want the user to perform.

The entire onboarding fits on a single page, without scrolling, on a web interface. You should not be too worried about this in a work context, but it is still best practice to keep the initial conversation above the fold.

Let's continue to the next step, the team onboarding script, shown after the bot is invited to a channel by the installer (Figure 17-20).

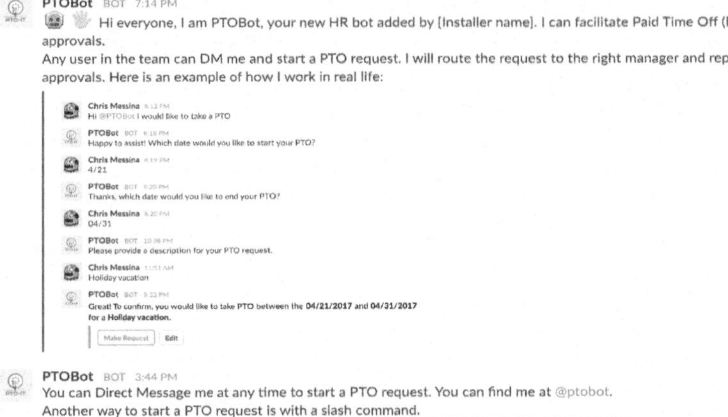

FIGURE 17-20.
The team onboarding script

The text is very similar to the installer script, but you will notice I have added a little text decoration at the end of the team onboarding text. I surrounded the slash command with backticks (``) to render the text as a code block. This hints to the user that the slash command is like a short command line that they can use, and that they should pay attention to the parameters the command accepts (in the same way one does when running a script on the command line).

Now, let's move on to the main flow. To remind you, the main functionality of our PTOBot is as follows:

1. Employee requests PTO in a direct message with the bot (or a slash command).
2. Manager gets a notification and approves/rejects the request.
3. Employee gets notification of approval/rejection.
4. Team gets notification of PTO.

This is by no means a simple "Hello World"–style process. I did not want to avoid complexity, but wanted to demonstrate the flexibility and unique attributes possible in bots for a work environment. We will design each step in this process, learning and improve each step along the way.

We'll design each of these steps in a separate Walkie flow, starting with the PTO request (Figure 17-21).

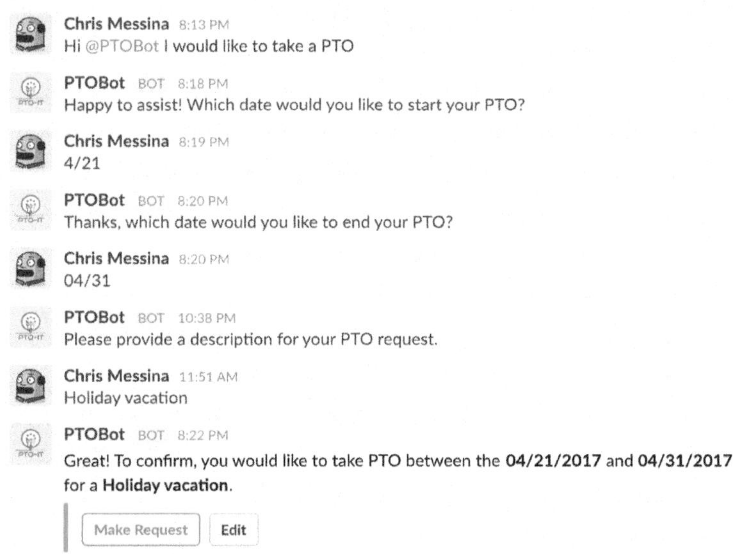

Chris Messina 8:13 PM
Hi @PTOBot I would like to take a PTO
PTOBot BOT 8:18 PM
Happy to assist! Which date would you like to start your PTO?
Chris Messina 8:19 PM
4/21
PTOBot BOT 8:20 PM
Thanks, which date would you like to end your PTO?
Chris Messina 8:20 PM
04/31
PTOBot BOT 10:38 PM
Please provide a description for your PTO request.
Chris Messina 11:51 AM
Holiday vacation
PTOBot BOT 8:22 PM
Great! To confirm, you would like to take PTO between the 04/21/2017 and 04/31/2017 for a Holiday vacation.
Make Request
Edit

FIGURE 17-21.
The PTO request script

You will notice that I have used some lightweight formatting by making the dates and the description captured stand out in bold (surrounding them with *s) and kept the green color coding for actions we would like the user to take.

As you can see, the conversation is long with a few places for potential errors. This is where our Slack command comes into play. Let's see the same conversation compacted to a couple of lines (Figure 17-22).

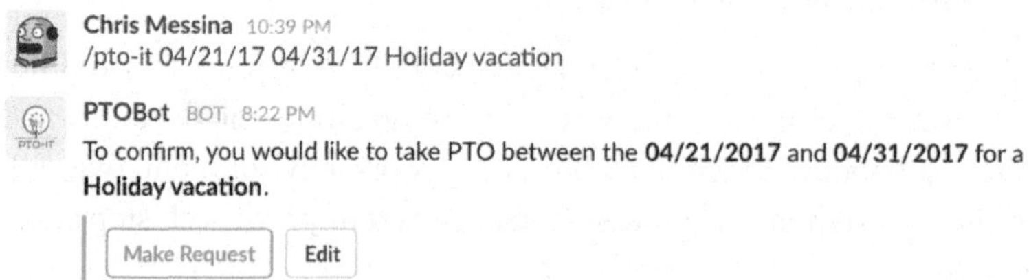

FIGURE 17-22.
Designing the slash command interaction

In a single line the user has provided all the necessary information to the bot, initiating a PTO request without the need for a lengthy conversation. Slash commands are great when you have a small and structured set of entities (variables) your bot needs to extract, and a savvy set of users who can remember how to use the commands.

Now, let us continue to the manager approval step (Figure 17-23).

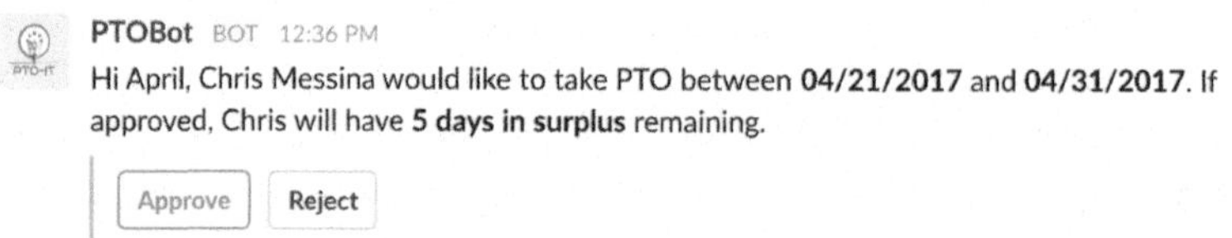

PTOBot BOT 12:36 PM
Hi April, Chris Messina would like to take PTO between 04/21/2017 and 04/31/2017. If approved, Chris will have 5 days in surplus remaining.
Approve
Reject

FIGURE 17-23.
The manager PTO approval script

This is OK, but it could be better. The name of the game here is *get things done as fast as possible*. This mean rendering the information in the easiest possible way to digest. We made the important parts bold, but I think the way the message is currently structured forces the user to read through it in order to get the necessary information. Let's see if we can enter the details in Slack's structured template (called a message attachment) and make it easier to digest and act upon. Figure 17-24 shows the result.

PTO-IT
PTOBot BOT 8:29 PM
Hi April - You have a PTO request waiting for your approval.

FIGURE 17-24.
Rendering the request details in a message attachment

I think this might be an easier way for the manager to pull out the relevant data. Of course, we will have to test it with actual users, as this is only an assumption.

Now let us finish up the approval step (Figure 17-25).

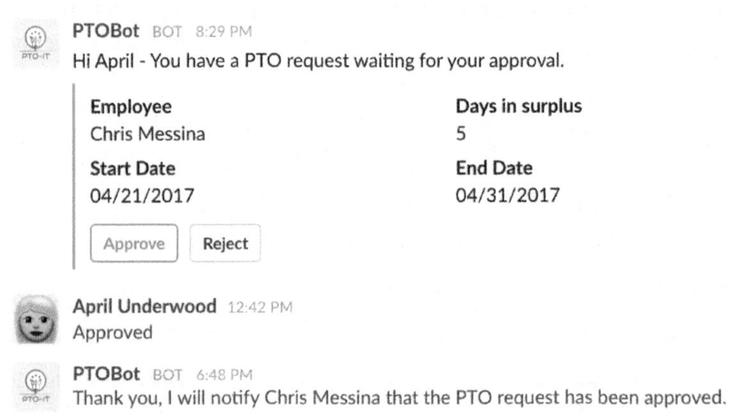

PTOBot BOT 8:29 PM
Hi April - You have a PTO request waiting for your approval.
Employee
Chris Messina
Days in surplus
5
Start Date
04/21/2017
End Date
04/31/2017
Approve
Reject
April Underwood 12:42 PM
Approved
PTOBot BOT 6:48 PM
Thank you, I will notify Chris Messina that the PTO request has been approved.

FIGURE 17-25.
Continuing the approval script

Now that we have designed the script, you might notice a few shortcomings with this design. The buttons are still there, and there is a chance the user will click on them by mistake. There is also a good chance that this design will be messy in a real-life scenario, when multiple requests might come in concurrently. It will be hard to manage the requests and keep track of which have been approved and which were rejected. Let's try another approach (Figure 17-26).

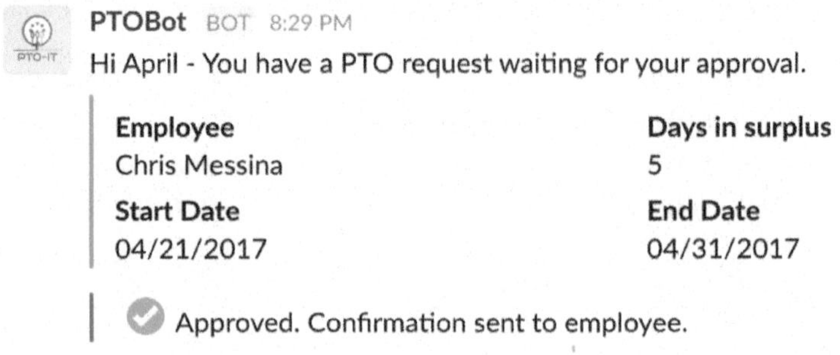

FIGURE 17-26.
Fine-tuning the approval script

In this design I replaced the buttons, once the user has clicked "Approve," with an approval confirmation. I think this is a better way to implement the process. Replacing the buttons ensures the user does not press one of them again by mistake. It also removes some of the cognitive load, if a lot of messages like this one appear in a conversation, and gives the user the feeling of accomplishment that users love in todo lists.

In the Walkie tool itself, I have forked the approval flow into "request approved" and "request rejected." Figure 17-27 shows what the "rejected" flow looks like.

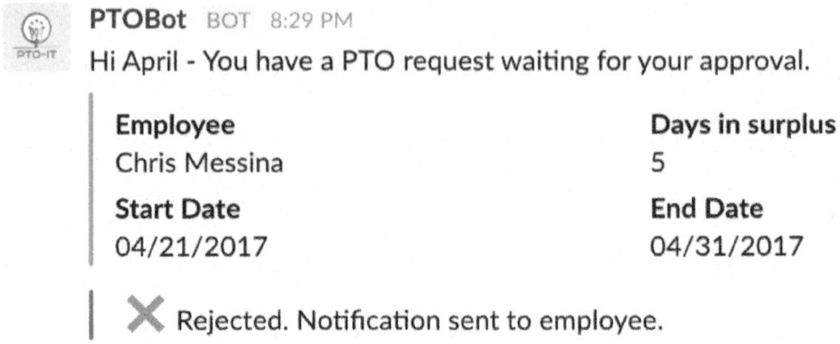

FIGURE 17-27.
The "request rejected" flow

Now it is easy to see at a glance which requests have been accepted or rejected, and the user does not need to read through a text conversation to see which requests have been handled. In more advanced versions we might want to add a reason for rejection, but let's keep it simple for now.

Next, in the employee notification flow, we will implement what we've learned about message formatting and button replacement (Figure 17-28).

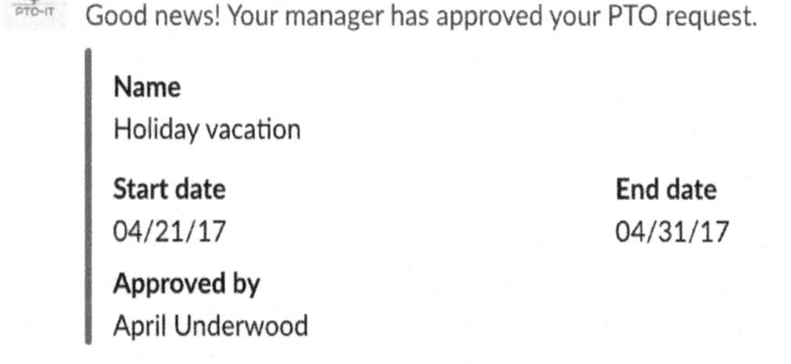

PTOBot BOT 7:48 PM
Good news! Your manager has approved your PTO request.
Name
Holiday vacation
Start date
04/21/17
End date
04/31/17
Approved by
April Underwood
Would you like me to send a notification in the #PTO channel?

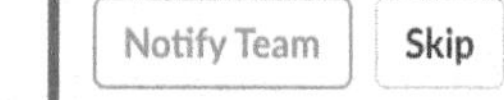

FIGURE 17-28.
The employee notification script

Notice that the bot is actually rendering two message attachments—one is informational, color coded in blue, and the other is actionable, color coded in green. Clicking on "Notify Team" will follow the same practice of replacing the buttons with a confirmation that the notification has been sent (Figure 17-29).

FIGURE 17-29.
Replacing the buttons with a confirmation

Note that we also changed the color coding of the second attachment to blue as it moved from actionable to informational. I chose to use this color schema as an example of how color coding can help with mental load reduction. We will have to test if this resonates with our users later on.

Let's finish this step by suggesting that the user install VacationBot (Figure 17-30).

PTOBot BOT 8:09 PM
If this PTO is a vacation, you might be interested in our new VacationBot offer.
This is a Facebook Messenger bot that can inform you about activities and corporate discounts at your destination.
Here is the link to install it - http://m.me/somelink

FIGURE 17-30.
Recommending VacationBot

This is a unique pattern of one bot *recommending* another bot to the user, and on a different platform. Clicking on the link will take the user straight to an onboarding conversation with VacationBot on Facebook Messenger.

Meanwhile, let's finish the main flow by designing the team PTO notification (Figure 17-31).

Employee
Chris Messina

Start date	**End date**
04/21/17	04/31/17

Approved by
April Underwood

FIGURE 17-31.
The team PTO notification

A PTO process like this is traditionally done manually, using paper forms, spreadsheets, or web tracking tools; it can be messy and require a lot of time. Our assumption is that users will find this process easier, more intuitive, and more productive.

The last thing to take care of is the logo. As mentioned earlier, particularly in our VacationBot, it looks small and indistinct. We also want the logo to be consistent in both bots. Moving forward, I will use the simple logo shown in Figure 17-32.

FIGURE 17-32.
The new logo for VacationBot and PTOBot

User Testing

Now that we have designed the main flow of both bots, it is time to put them in front of actual users. First, we need to decide how we want to test our design.

There are a few options:

1. Show users a video or a step-by-step replay of the conversation and get their inputs.
2. Create a mock (fake) bot and let users play with it.
3. Create a working alpha and let users work with it.

Both Botsociety and Walkie support replaying the conversation either as a movie or a step-by-step walkthrough. Showing potential clients/users these videos can get you very valuable feedback. You will not be able to see users perform tasks themselves, which might be the most important indication of good design, but you will get feedback fast and with little development cost.

As for mocking a bot, Walkie goes a step further and lets you export the conversations into a JSON-format file. An engineer can plug this file into a script that mimics an actual bot. A mock bot is a great tool for testing interactions, and it doesn't really matter that the bot is not connected to the real backend system. In our case, we don't care that the bot is not connected to a real PTO system, or that the data is fake. Getting a user to go through the request process and a manager to go through the approval process in real life will surely teach us a lot.

[KEY TAKEAWAY]

A mock bot is a great tool for testing interactions. It doesn't really matter that the bot is not connected to the real backend system.

Alternatively, if you are confident with your design, you might even start building the actual bot and get feedback from live alpha users. This is useful because it is the shortest path to production, if you get it right. Users test the real bot, with real data, and you get live and super-accurate feedback. This option will work well if you have clients who are willing and able to be your alpha users and use your software in real life.

Laura Klein has written a great "Step-by-Step Usability Testing Guide" (*https://guides.co/g/usability-testing-guide/7996*) that outlines the steps in usability testing. Assuming that we will create a mock that users can play with, let's discuss the usability testing steps for our PTOBot.

BEFORE YOU START—PROTOTYPING A MOCKUP BOT

First let's create a mockup of our PTOBot. In order to prototype simple processes we will use a tool I developed called ProtoBot, which you can install freely by searching for *ProtoBot* in the Slack app directory (*https://slack.com/apps*).

ProtoBot does not require coding skills, and it is really easy to create mockups of bots with it. You install ProtoBot in a testing team of your choice, and start a conversation with it. ProtoBot provides a detailed description of how to use it, but we'll go through a short example here.

ProtoBot can mimic multiple bots—that is why I initially called it "Dr Jekyl." These bots are called *personas*, and ProtoBot can assume a persona with the following steps.

In a direct message with ProbotBot:

1. Type */new-persona PTOBot* to start a new bot persona script.
2. Type */set-persona-name PTOBot* to set the name the bot will use in this script.
3. Type */set-persona-icon-url [URL]* to set the icon the bot will use in this script.
4. Say hello to your new bot (Figure 17-33).

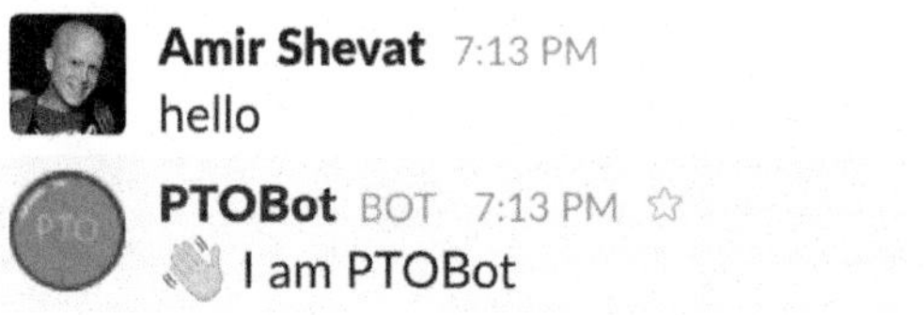

Amir Shevat 7:13 PM
hello

PTOBot BOT 7:13 PM
I am PTOBot

FIGURE 17-33.
ProtoBot learning to be PTOBot

The way to teach ProtoBot a new script is simple: you just start talking to the bot and follow its instructions (Figure 17-34).

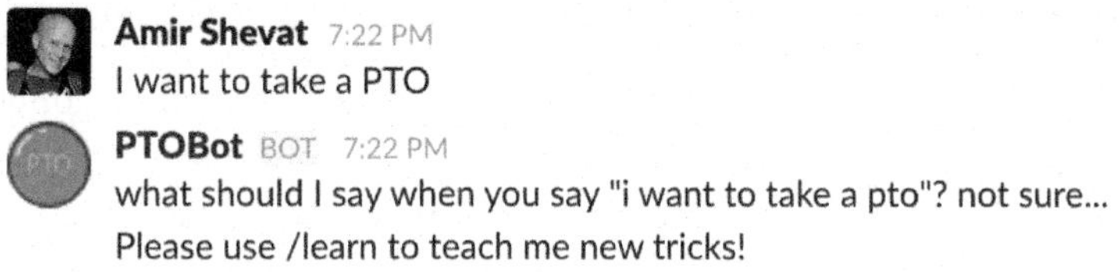

Amir Shevat 7:22 PM
I want to take a PTO

PTOBot BOT 7:22 PM
what should I say when you say "i want to take a pto"? not sure...
Please use /learn to teach me new tricks!

FIGURE 17-34.
ProtoBot tells you how to teach it a reply

/learn is a slash command you can use to teach ProtoBot what to say when the user says something. It follows this pattern:

/learn [user says]

[bot says]

Note the newline between what the user says and what the bot says (use Shift+Enter to create this newline in Slack). Let's teach PTOBot what the script replies when the user says "I want to take a PTO" (Figure 17-35).

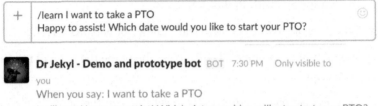

FIGURE 17-35.
Teaching the bot a new script

Now, after we have trained ProtoBot with this step in the script, let's run the same script again (Figure 17-36).

FIGURE 17-36.
The bot has learned the correct reply

Yes! ProtoBot is starting to learn how to mimic the PTOBot persona.

In this way, using the */learn* slash command, you can teach ProtoBot the entire script. It is important to note that ProtoBot does very little pattern matching, when it comes to user inputs—it is case insensitive and ignores characters like question marks or periods, but you will need to teach it the various permutations of possible user inputs for a given script. In this example, for instance, you can also teach the bot to handle inputs like "I want PTO" or "Start PTO request" to start the PTO request script.

ProtoBot has more advanced functionality, such as the ability to import JSON files from tools such as Walkie and support for multiple concurrent personas, but you can read all about it by just saying "help" to the bot in a direct message at any time. ProtoBot is a good tool for basic mockups, but it does not support complex scripts like contextual help. If you want to implement these, you might want to code your mockup or find a tool that supports these advanced features.

Once we have taught ProtoBot a section of the script, we can start thinking about the next steps in our usability testing.

PLANNING THE TEST

To test PTOBot we would like to find existing PTO-IT users—preferably friendly ones and early adopters—and invite them to PTO-IT's office to do this experiment. We will also invite a few non-clients to see if VacationBot can be used as a standalone product.

We would like to invite users who are already using Slack, because we do not want to have to teach them how to direct message or how a channel works during the usability testing.

CREATING TASKS AND DISCUSSION GUIDES

We will focus on two tasks: PTO requests and PTO approvals. We will create an overview of our PTOBot and collect responses to the following background questions from our participants:

1. Managers:
    a. How many people do you manage?
    b. How many PTO requests do you handle in a month?
    c. What system or tool do you use for this today?
    d. Do you use bots for your day-to-day tasks?
    e. Do you use bots/Slack mainly on the web or on mobile?
2. Employees:
    a. How many PTO requests do you make in a year?
    b. Who approves your PTO requests?
    c. What system or tool do you use for this today?
    d. Do you use bots for your day-to-day tasks?

e. Do you use bots/Slack mainly on the web or on mobile?

f. How do you let your team know you are taking PTO?

We will also create a task for each participant:

Manager
: You are a busy exec with little time for paperwork. Show me how you use PTOBot to quickly get through your PTO approval tasks and get back to what's important.

Employee
: You want to take a vacation to Cancun for 10 days starting 04/21/17. Use PTOBot to get your paperwork in order.

RECRUITING PARTICIPANTS

We will recruit four managers and four employees. Half of them should be current PTO-IT users and half of them will be new potential users who have not worked with PTO-IT in the past.

SETTING UP THE ENVIRONMENT

We will set up a clean computer with Slack installed and a test team setup. We will add ProtoBot to that team and teach it the PTO request script and the PTO approval script. Note that if it is easier for you to mock up the bot using any other tool, that's fine too!

You can also set up the environment by opening Slack and showing the onboarding message from the bot in the HR channel, as that is the most likely place that users will come to learn how to use the bot.

MODERATING THE SESSIONS

Make sure to explain that this is just a prototype, and might not work perfectly or look final. Let each user try to perform the assigned task, and possibly fail, without interfering. Make sure to train ProtoBot well beforehand so that it can handle many permutations of user inputs. Also make sure to support the help script so that the user can get help if needed.

Make sure the users are comfortable and know it is OK to fail in these tasks. Ask them to speak their thoughts out loud while performing the task, and follow up with questions—for example, when a user makes a statement like "That was easy," ask "What was easy about it?" Keep the same pattern for all types of comments. Take notes on all actions and comments made by the users.

ANALYZING THE DATA

Mark each user test with "Task Completed," "Task Completed with Difficulty," or "Task Failed." Correlate your notes on the comments and find patterns that signal issues. Here are some examples:

1. Two employees had a hard time entering the start date of their PTO in a way the bot can understand.
2. Three managers thought the PTO approval process was "very easy."
3. One employee started a PTO request in the channel rather than in a direct message.

IMPROVING AND ITERATING

Order the problems you identify by frequency of them happening and their severity, and start fixing these issues. For the examples we outlined above, here are the correlated fixes:

1. Two employees had a hard time entering the start date of their PTO in a way the bot can understand. *Support multiple ways to enter a date. Use AI solutions if necessary.*
2. One employee started a PTO request in the channel rather than in a direct message. *Create an error script that lets the user know that PTO requests only work in direct messages, and guide the users to start a direct message.*

Once you've fixed these issues, run the test again with fresh test users, and enjoy the fact that you are making your bot better with each iteration.

This was an example of running a usability test to learn about and improve your bot before you launch it. There might be other types of tests you would like to run; ones that check user satisfaction or brand impact, for example. You might also use a different tool to prototype or create an alpha of your bot in order to run these tests. The most important thing is to iterate and learn.

Learning and iterating does not stop when you launch your bot in production—quite the opposite. Most bot builders I've spoken with have told me they have learned the most from looking at users using their bots in production, and collecting logs and analytics that they can use to constantly improve their bots. This will be the topic of our next chapter.

[18]

Bot Building Overview

Whatever good things we build end up building us.

—JIM ROHN

BUILDING A BOT IS a topic worthy of a book of its own; it would be foolish of me to presume to teach anyone to build a bot in one chapter. Moreover, this is a design-focused book, so it might not be relevant to you to know all the details and processes involved in engineering a production-ready bot. Having said that, it is important for everyone in this industry to have a basic knowledge of how the technology works and what kinds of tools you can use to build a bot.

Bot Architecture

In contrast to mobile apps, bots are not installed (deployed as native code) on the platforms on which they run. Instead, the bot connects to the user or team through a set of application programming interfaces (APIs). In addition, the bot service does not ever communicate with the chat client directly; it is all done through the proxy of the chat platform service provider (e.g., Slack, Facebook, Kik, or Amazon's servers).

[KEY TAKEAWAY]

Bots are not installed in the messaging clients; they are connected to the clients via APIs.

As you can see in Figure 18-1, the chat platform relays messages from the clients to your bot, and can add information such as presence and other events. The chat platform lets your bot send messages and perform additional operations, such as creating channels, inviting users to a conversation, and broadcasting messages. Each platform has a slightly different API, depending on the platform's capabilities. This architecture keeps the chat client secure and provides the bot builder a clean interface to connect to (with the exception of webviews, supported by some platforms). The customer can be on iOS or Android, macOS, or even Linux; the bot builder does not need to care about the client-side implementation. This also implies that the bot is hosted on the bot builder's servers or by a cloud service provider.

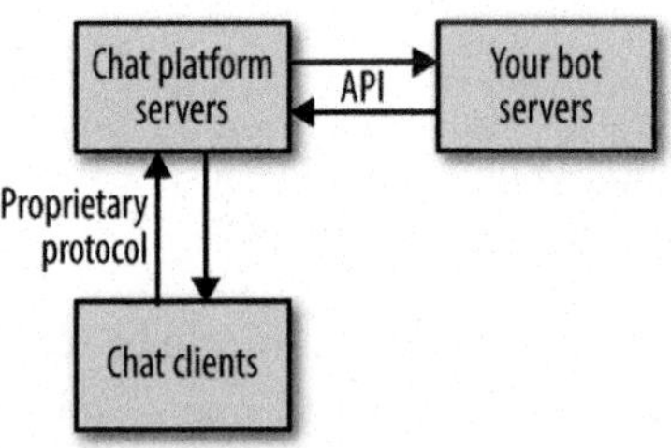

FIGURE 18-1.
The chat platform connects chat clients to your bot's servers via APIs

Bot Building Technologies

Bot builders have several tools at their disposal. Some require more technical skills than others, while providing higher levels of flexibility. Here are several examples of tools bot builders use, from the least technical to the most.

VISUAL AUTHORING TOOLS AND INTEGRATED DEVELOPMENT ENVIRONMENTS (IDES)

Tools in this category provide you with an integrated development experience—they're visual authoring tools, usually with AI integration, that will even host your bot. The key value proposition of these tools is the ease of getting started, the speed to market, and the ability to get a bot running with minimal technical skills.

Flow XO (*https://flowxo.com*) is a good example of an easy-to-use authoring tool that is cross-platform and provides you with a rich set of integrations and predefined templated flows, such as Small Talk, that can be added to your bot (Figure 18-2).

Flow XO's biggest value proposition is a set of more than 100 integrations you can add to your bot, including to services such as email, Google Calendar, Google Docs, Stripe, and more.

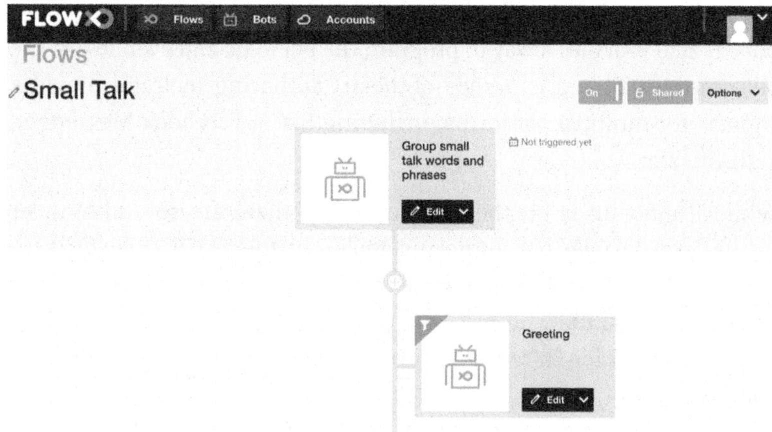

FIGURE 18-2.
The Small Talk template in Flow XO

PullString (*https://www.pullstring.com*) is an excellent example of an advanced authoring tool. The PullString team actually started creating conversational interfaces way before the bot era. They build the interfaces for some of our favorite toys, providing toy manufacturers with a tool to script the toy conversations. Today, they have ported their solution to a bot-focused integrated development environment that runs on your computer (Figure 18-3).

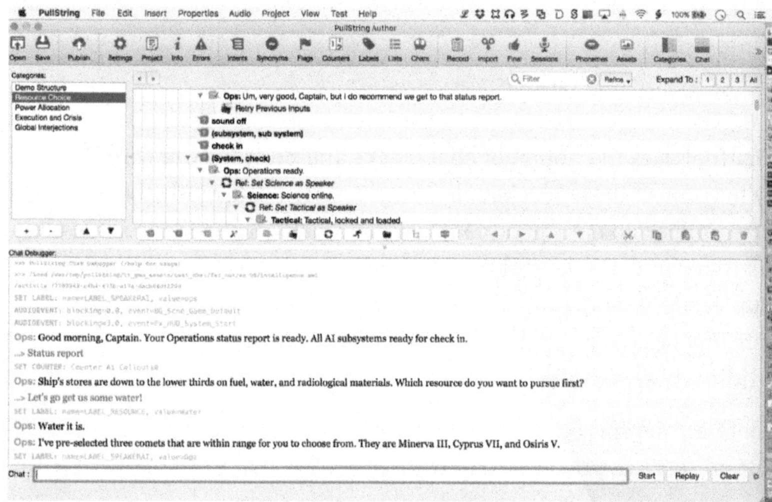

FIGURE 18-3.
The PullString IDE

PullString provides bot builders with a rich interface to script their bots. It also provides a way to program the bot logic and even to test the outcome. PullString is a state-of-the-art authoring tool and provides support for multiple platforms, including Slack, Facebook Messenger, and Amazon.

While PullString is on the advanced side, there are several simpler (while less flexible) bot authoring tools out there. Chatfuel (*https://chatfuel.com*) is a popular example of a simple bot authoring tool. With Chatfuel, you actually do not need to download anything to your computer—you just log in to the Chatfuel website and start building your bot (Figure 18-4).

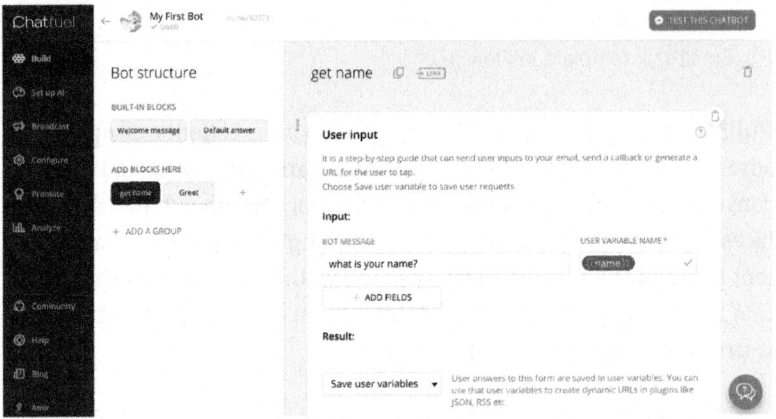

FIGURE 18-4.
Getting started with Chatfuel

Chatfuel uses the notion of chat blocks and provides you with the ability to connect these blocks into a conversation. Currently the only platform supported is Facebook Messenger. Chatfuel provides a set of plug-ins that can cover several common use cases (set user variables, go to plug-in, Zapier integration, send to email, etc.), but if your bot has custom business logic that part of the logic will need to be hosted on a third-party server.

There are several other tools, at different levels of maturity, as well as several services providers. Here are just a few that I have worked with:

Chatflow (https://chatflow.kitt.ai)
A promising beta service that lets you build your conversational interface while focusing on the flows. It supports multiple platforms, including Slack, Kik, Amazon, and Facebook Messenger.

Pandorabots (https://www.pandorabots.com)
Pandorabots is one of the oldest web services for building and deploying chatbots out there. Pandorabots takes an XML approach to authoring and provides bot builders with the ability to define their conversations through XML, as well as hosting and running their bots. Pandorabots supports multiple platforms, including Slack, Kik, Amazon, and Facebook Messenger.

Automat (http://www.automat.ai)
Automat is a platform designed to bridge human expertise and artificial intelligence to make it easier for anyone to build a conversational bot. Automat is in private beta at the time of writing, but it powers some popular Kik bots, such as the CoverGirl bot.

Recast.AI (https://recast.ai)
A flow-focused authoring tool that supports various platforms, including Kik, Skype, Slack, Facebook Messenger, and even email.

Imperson (http://imperson.com)
Marketing itself as the "WordPress for bots," Imperson provides a friendly authoring tool together with a hosting solution.

This is not a comprehensive list; there are many other great tools out there, and a lot of new ones are sure to emerge.

Similar to web- and mobile-focused no-coding tools, authoring tools like these are a great way to get to market fast without a steep learning curve. As with all tools, there is a balance of ease of use and flexibility (as well as ability to deal with complexities), and bot builders need to evaluate these compared to other options outlined in this chapter, such as SDKs and roll-your-own solutions. You can explore these tools for your prototyping stage, as most of them are really ease to use, and also for the production stage if they meet your full requirements.

ARTIFICIAL INTELLIGENCE (AI) SERVICES

There are several AI services that provide you with the ability to define the conversation in a textual or visual way, and let the AI manage and handle the runtime conversation with the user. Some of these AI services can host your bot or call an API that provides AI processing as a service. Here are some popular AI services:

Google API.AI (https://api.ai)
: API.AI was a pre-bot conversational solution that powered Android and iOS personal assistant apps. It now focuses on providing an easy-to-use and simple AI service that can be used to build bots for platforms such as Slack and Facebook Messenger. API.AI was acquired by Google in 2016 and now provides AI services to many platforms, including the emerging Google Home (a product competing with the Amazon Echo). Google provides several other AI solutions, such as its Vision API and TensorFlow, as part of its Cloud Platform.

IBM Watson (https://www.ibm.com/watson/)
: IBM Watson is a comprehensive AI toolset that offers everything from conversation management to sentiment and image analysis. One of Watson's strengths is its support for exporting and self-hosting your AI model. Watson is popular with enterprise bot solutions.

Facebook Wit.ai (https://wit.ai)
: Wit.AI is a conversational AI solution provided by Facebook. While Facebook is a chat platform of its own, Wit.AI is not part of that platform and provides services that can support other chat platforms.

Microsoft LUIS (https://www.luis.ai)
: Microsoft released its Language Understanding Intelligent Service (LUIS) in 2016 with the aim to compete in the AI market. LUIS is a standalone service that supports multiple chat platforms and integrates well with the Microsoft Bot Framework (discussed in the next section).

msg.ai (http://msg.ai)
: msg.ai is an early-stage AI solution focusing on conversational commerce use cases.

Artificial intelligence is not a mandatory part of all bots, so the first step is to decide whether or not you need an AI service. If you think your bot requires an AI solution, you should evaluate a few of these options based on your requirements, supported features, ease of use, and service level.

[**KEY TAKEAWAY**]

Artificial intelligence is not a mandatory part of all bots, so the first step is to decide whether or not you need an AI service.

SOFTWARE DEVELOPMENT KITS AND BOT FRAMEWORKS

Software development kits (SDKs) and bot frameworks are software components that wrap the messaging platform APIs and provide bot builders with an easier and programmatic interface for building their bots. Here are some common SDKs and frameworks:

Botkit (https://www.botkit.ai)

Botkit is the most popular bot framework currently out there. Botkit is open source, extendable, and cross-platform. The team that maintain Botkit are bot builders themselves and run a successful bot called Howdy. They are also active members of the bot community. The team are now experimenting with a simple authoring tool that integrates with Botkit and lets bot builders decouple authoring from business logic implementation.

Microsoft Bot Framework (https://dev.botframework.com)

The Microsoft Bot Framework is an offering that you can use to build and deploy bots. The framework consists of the Bot Builder SDK, Bot Connector, Developer Portal, and Bot Directory. There's also an emulator that you can use to test your bot. The framework supports multiple chat platforms, such as Slack, Kik, Telegram, and Facebook Messenger, and integrates well with LUIS and the Azure cloud offering.

Slapp (https://github.com/BeepBoopHQ/slapp)

Slapp is an easy event-driven SDK for Slack bots. It integrates well with the company's hosting solution, called Beep Boop, which we will discuss later in this chapter.

Twilio SDK (https://www.twilio.com)
: If you are interested in SMS bots, then the Twilio SDK is what you need. It provides a simple and easy-to-use framework for building bots that communicate via text messages.

SDKs are a good solution if you are comfortable coding your bot by yourself but do not want to deal with the plumbing. An SDK provides an easier-to-use development experience by dealing with the complexities and edge cases of the messaging platform API. The disadvantages of SDKs are some loss of flexibility and dependency on the SDK for updates and new feature support.

ROLL YOUR OWN

A very common solution for professional bot builders is to build your bot using plain coding skills and the messaging platform APIs. The advantage of this approach is absolute flexibility and control over your bot and its flow; the disadvantages are a steep learning curve, longer time to market, slower development cycles, and dependency on engineering resources.

HOSTING SOLUTIONS

As bots are hosted services that connect to messaging platforms, they need to be hosted on internet-accessible servers. There are a few options for this:

- *Use the hosting provided by your tool provider.* For example, IBM Watson can also host your bot on its servers.
- *Use any cloud hosting provider.* Most cloud services, such as AWS, Azure, Google Cloud, and Heroku, provide infrastructure you can use to host your bot.
- *Use your own servers.* You can run your bot on your own servers, as long as they are accessible through the web. Some of my early-stage bots are hosted on a $1/month PHP shared server.
- *Use Beep Boop.* Beep Boop is a hosting service dedicated to bot hosting. It provides you with a management console that is integrated into Slack. You can basically manage your bot with the Beep Boop bot using chatops (Figure 18-5).

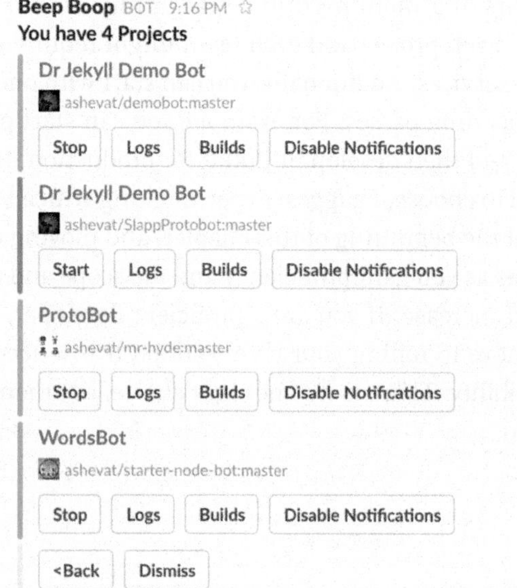

FIGURE 18-5.
Beep Boop integrates well with the Slapp and Botkit frameworks and currently supports Slack bots

Picking the Right Tool

Which tool is the right tool is almost a theological question. There are many parameters that need to be taken into account:

- Technical competency of the team
- Time to market
- Complexity of the bot
- Integrations with other systems
- Availability of resources
- Cost
- Security
- Company policy
- Personal preferences

All of these parameters and more need to be taken into account and weighted accordingly. Each project and each team might require a different set of tools and services. Additionally, you can start with one tool and move to another as time passes. For example, you can start prototyping with Flow XO and then develop in Botkit for production. If you do not know what tool to choose, I suggest experimenting with a few of the easy ones (listed at the beginning of this chapter) and moving on to the more complex ones as you gain proficiency and the scope and complexity of your project increase. If you are a proficient developer, start with the SDKs and move to rolling your own solution if you need the extra control and flexibility. There is no one right choice, but there is a lot of exploration to do.

[19]

Analytics and Continuous Improvement

If you torture the data long enough, it will confess anything.

—GORDON TULLOCK

IF YOU'VE EVER MANAGED a website or a mobile app, you probably got addicted to checking your analytics—waking up every morning and seeing that you got more users and more engagement is a delightful experience. It can also be an eye-opening experience to see how users engage with your service, which buttons they click, and which parts they avoid. Great products and great design stems from understanding your users, and data from analytics services about how people really use your bot should be treated like gold.

How Do Bot Analytics Work?

Bots connect to a messaging platform (like Kik, Slack, Facebook Messenger, and others) through an API that lets you send and receive messages, receive events such as being invited to a direct or team conversation, receive presence change events, and so on. When users use your bot, you will be getting a great deal of data about the interaction, and the key challenge is to turn this raw data into insights you can use to understand your users and improve your bot. There are a few common ways you can turn that data into actionable insights:

Raw logs

Many bot builders spend lots of time looking at the raw data, seeing what users are saying to their bot and what the bot is saying back. This approach is time-consuming, but provides an unfiltered view of the conversations your bot is having. This might be easiest way to start, although the other options are pretty easy to implement.

Filtered logs

Whenever something significant happens, log it as important and then view and analyze your logs using a tool that can filter log entries. A good example of events bot builders usually log is successful and unsuccessful outcomes of conversations—if your user has bought an item, log it; if your user has failed to buy an item, log it. It is important to log relevant information with the event. For example, if your user has failed, log the place in the conversation where the failure happened, and correlate it with other logged events to see if this is a consistent issue.

Analytical tools and SDKs

If you want to see trends, and more indicative insights from a growing data stream, you might want to consider integrating your bot with an analytical tool or SDK. These tools distill the logs, or a stream of raw information, into insights. They provide data visualizations and flag key trends. You can build your own analytical tool, but there are a lot of good ones out there.

SaaS analytical services

These are very similar to the previous category, but are provided to you as a hosted third-party service. You integrate a few lines of code into your bot, and these lines of code send a copy of the data to the third-party service. The service in turn provides you with a dashboard of visualized insights. Google Analytics is a great example of a hosted service for website analytics, and we will review some bot services later in this chapter. The benefit of a hosted solution is ease of use. The downside is that you share this information with a third party—something that is prohibited in some sensitive use cases.

These methods are not mutually exclusive. I recommend using a combination of the first two, and either the third or fourth:

- Use *raw logs* at the start to see a detailed stream of your bot conversations and collected data.
- Use *filtered logs* on an ongoing basis to flag errors and positive significant outcomes.
- Use an internal or third-party hosted *analytics solution* to get aggregated insights and trends.

Looking at your data and optimizing your bot is critical. There are a few ways to do that, but the important thing is that you actually do it. Next, we will see examples of how logs and analytics solutions can drive better bot design.

[KEY TAKEAWAY]

Looking at your data and optimizing your bot is critical. There are a few ways to do that, but the important thing is that you actually do it.

Looking at Logs

When the Statsbot team initially designed their bot, they thought users would like to have natural language conversations with the bot. The assumption was that most users are explorers and would like to access the marketing insights using an interactive and conversational approach.

The team assumed this type of conversation:

> *User*: @statsbot: how many users last month
>
> ***Statsbot***: There were 1240 users last month. Do you want to segment users by country or source?
>
> *User*: By country
>
> ***Statsbot***: US: 900
>
> ***Statsbot***: France: 140
>
> ***Statsbot***: Germany: 100
>
> ***Statsbot***: Other: 100
>
> ***Statsbot***: Say "mobile" or "new" to filter results to only mobile or new users.
>
> *User*: Mobile
>
> ***Statsbot***: Out of 1240, there were 153 mobile users.

Looking at their logs, however, they soon discovered that their users knew exactly what information they were seeking, and were trying to take the shortest path and shorthand the conversation. So, they would see inputs like the following:

User: Users last month by country on mobile

The team then pivoted, making the design much more concise and focused on getting the users the insights they know they need, directly and in the fastest way possible. They deemphasized the natural text-based exploratory conversations.

According to Alyx Baldwin and Rachel Law, the cofounders of Kip, looking at logs proved to be one of the most useful things the team did to improve their bot's design. They have actually built their own tool to enable them to view the conversations the Kip bot was having with its users (Figure 19-1).

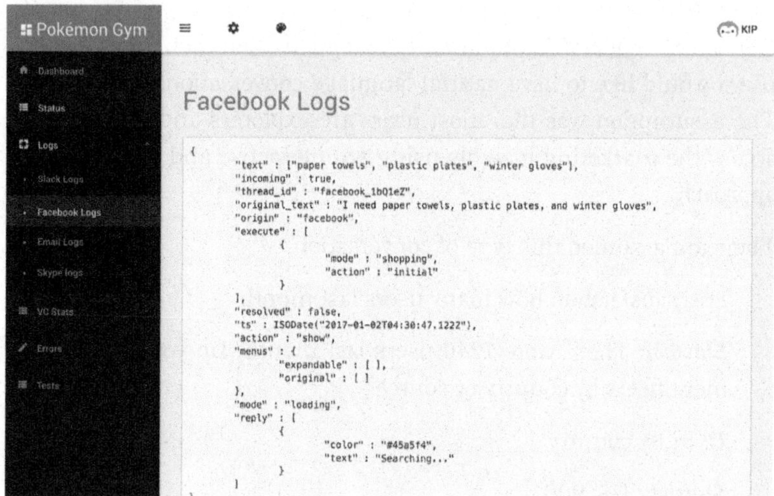

FIGURE 19-1.

Kip's tool for exploring the bot's logs

Insights from Analytics Solutions

As mentioned previously, analytics solutions fall into two categories: you can use a third-party hosted solution or a self-hosted analytics tool.

THIRD-PARTY HOSTED SOLUTIONS

There are numerous third-party options available to bot builders. We'll focus on one, to give you an idea of the capabilities such a tool can offer, but all of them are good; feel free to pick any that meets your requirements.

Dashbot (*https://www.dashbot.io*) is a bot-centric hosted analytics solution that is very easy to integrate into any bot. Once you add Dashbot's integration code into your bot, you can log into the Dashbot website and start getting aggregated insights about your bot.

At a high level, you can see how many users you have and their level of engagement (Figure 19-2).

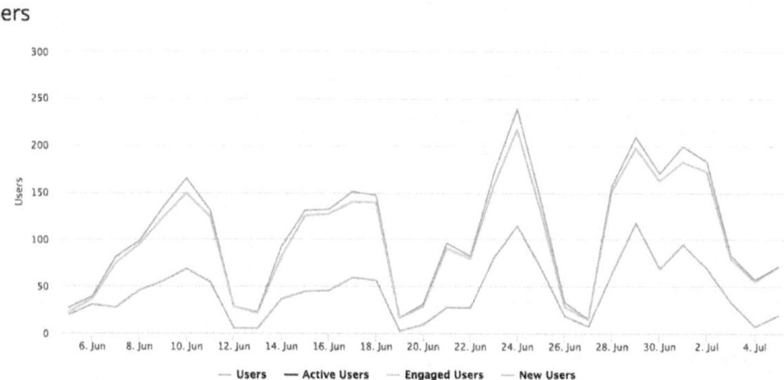

FIGURE 19-2.
Dashbot reporting on number of users and level of engagement

You can also see which languages your users are using (Figure 19-3).

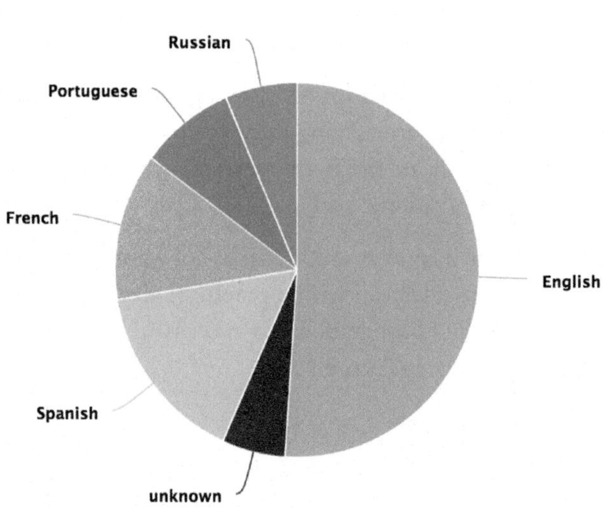

FIGURE 19-3.
Breaking down users by language

This can be significant. If you see a large portion of your users using a particular language, you might want to extend your support for that language.

A very hard thing to do when looking at logs is to see the path users take through the conversational interface. With analytical solutions, you can visualize the conversational path at an aggregated level (Figure 19-4).

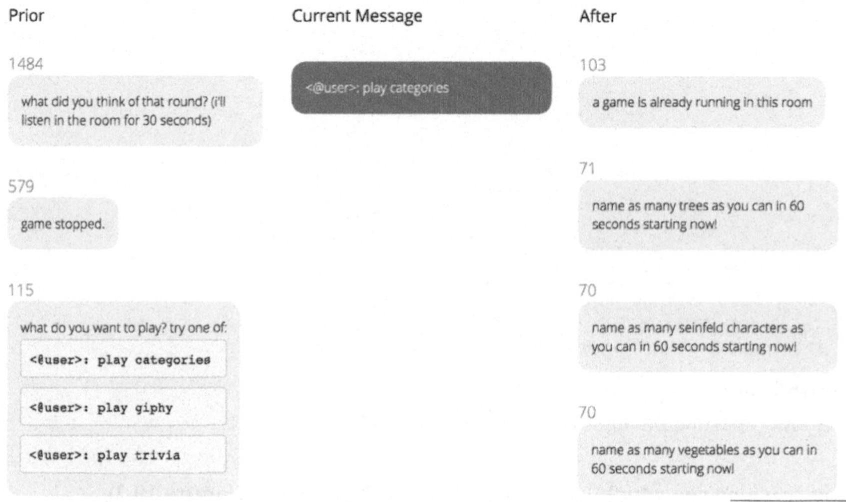
Prior
Current Message
After
1484
what did you think of that round? (i'll listen in the room for 30 seconds)
579
game stopped.
115
what do you want to play? try one of:
<@user>: play categories
<@user>: play giphy
<@user>: play trivia
<@user>: play categories
103
a game is already running in this room
71
name as many trees as you can in 60 seconds starting now!
70
name as many seinfeld characters as you can in 60 seconds starting now!
70
name as many vegetables as you can in 60 seconds starting now!

FIGURE 19-4.
Dashbot allows you to see what led to a particular user input, and what resulted from that input

In this example, you can see what led the user to saying "play categories" and what resulted from the user saying that. The numbers associated with each piece of text indicate the frequency with which that text has been shown in this conversational path.

If you want to deep-dive into a conversation, but do not want to look at your server logs, Dashbot provides you with a graphical view of all the conversations. In this example I have pulled in a successful interaction between my WordsBot and a user (Figure 19-5).

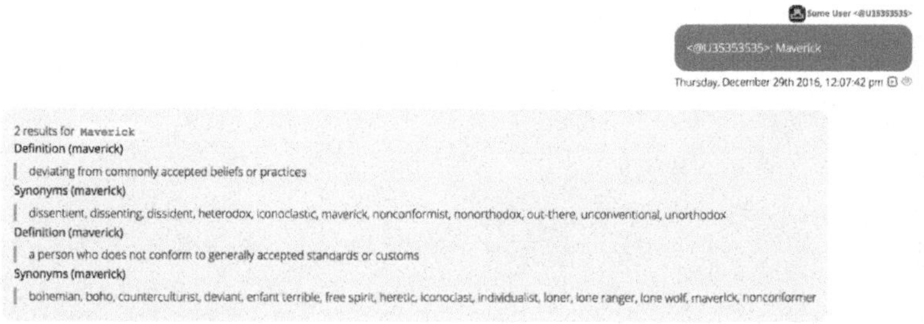

FIGURE 19-5.
Viewing a conversation in Dashbot

The tool provides you with information such as a transcript of the conversation itself, the time at which it occurred, and information about the user with whom the bot conversed. I have anonymized the example shown here to keep the user anonymous.

Dennis Yang, cofounder of Dashbot, explained to me why they added this feature:

> One of the biggest ways that we have learned about our bot was simply [by reading] its transcripts. Unlike on web or mobile, where you're trying to guess what users are trying to achieve by a series of events, conversational UIs have a perfect transcript of the session with your users. So, you no longer have to spend time and money setting up a user lab to watch your users interact; you literally have an exact description of what they are trying to do. Reading through transcripts is an invaluable way to gather subjective feedback about the usage of your bot.

These are just a few examples of the kinds of reports Dashbot and other analytical tools can generate. You will need to find the reports that are most significant to your bot business and derive your insights from there.

SELF-HOSTED ANALYTICS SOLUTIONS/SDKS

If you do not want to share your analytics data with third-party services, you can host it yourself. Botmetrics (*https://www.getbotmetrics.com*) is a self-hosting solution that you can integrate your bot's code with and host on your own servers.

One dashboard provided by Botmetrics that I really find useful, from a design point of view, is their *path* report (Figure 19-6).

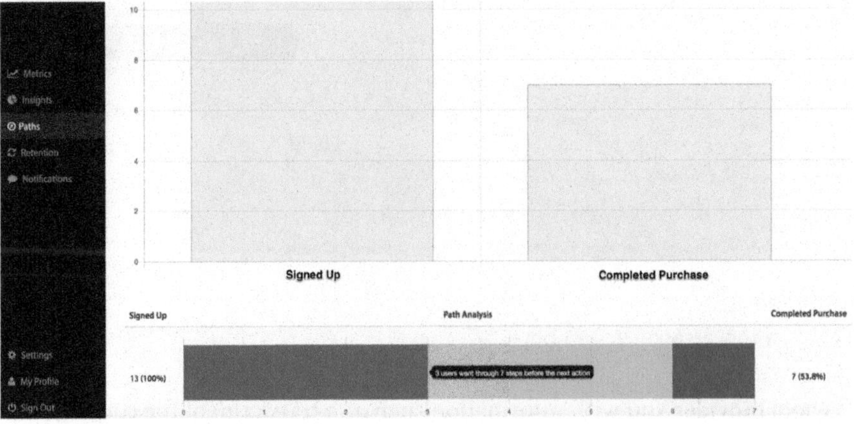

FIGURE 19-6.
Botmetrics path report

You can clearly see the conversion path between users signing up and completing a purchase in this example. You can also see insights such as the number of steps required in the path. With this report, you can experiment with different paths and correlate better conversion rates with different conversational design choices. This is a different way of looking at the user journey than Dashbot's reports: it looks at the conversational conversion rather than focusing on one step in the conversation.

Botmetrics also provides the source of its bot analytics tools, so users can extend it. This is a great solution if you want to build custom reports and extensions based on your bot's logic, or if you handle sensitive information and want to get analytical reports without sharing your data outside your servers.

BE CAREFUL NOT TO WORRY ABOUT THE WRONG THING

Many bot builders I meet report that they sometimes worry about the wrong metric. As an example, Figure 19-7 is a chart of user engagement with my WordsBot over a two-week period.

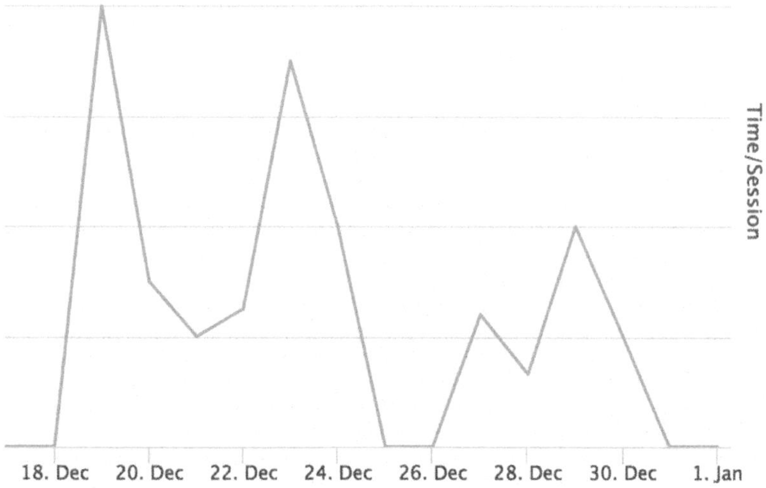

FIGURE 19-7.
WordsBot engagement metrics

You will notice that I have days with close to no engagement. This could make me worry and start wondering what is wrong with my bot. But this is actually a very common pattern with bots for work environments—there is very little to no engagement during weekends.

Another type of misguided worry is about fringe use cases. When your bot starts getting good traction you will start to see a rise in what seasoned entrepreneurs call "noise": users asking for, or complaining about, small things that are important to them but not indicative of the concerns of the general population. You will need to focus first on the "signals"—the important and common problems that most of your users have—and only then address the edge-case issues.

Continuous Improvement

If you have been working for more than a few minutes in the software industry, you know that software evolves all the time. We fix bugs, add features, redesign UXs, and run experiments. This is particularly important when it comes to conversations, as there is an expectation of "freshness" in our day-to-day conversations. Especially in consumer use cases, if we hear the same phrases again and again, we become tired of them.

I talked to Greg Leuch—head of product at Poncho, makers of one of the earliest and most successful consumer bots in the market—about their ongoing bot improvement cycles. Here is what he had to say:

> Our own internal tools include a bot conversation management system to allow editors to make live updates to Poncho, a broadcast and follow-on management system that provides custom targeting to users based on user profiles and preferences. We also have a bot conversation analysis tool that looks at all of our conversation histories within the scope of a network map, which will identify bounce rates, simulate probable conversation changes, and allow us to understand where we can improve conversation and dialogue.
>
> Our ability to retain users has been tremendously important. This is attributed to our constant A/B testing and conversation evaluation. Our bot analyzer has allowed us to identify and visually map our main conversation triggers. We post all of our unmatched (wildcard) responses into a Slack channel for our editors to identify ways to improve or extend conversations. And along with A/B testing, we challenge our teams to focus on retention, which helps build experiences and conversations that stick more for our users. We've been surprised at what some users want to talk about, and even more amazed when we build conversations with those users.
>
> If you are designing a bot it is important that you continue to observe and experiment. Always see where you can improve your conversations, add triggers for different phrases and misspellings, and try to lead the user toward solutions through conversation.

There are three recommended paths to improve your bot:

1. *Error reduction*—Find all the places the bot had a conversational failure and fix these instances. You can trigger logging of such errors every time an error script starts.
2. *Conversation optimization*—A/B test to reduce the steps in task-led conversations. Try to improve duration in topic-led conversations. Experiment with improving engagement and retention in both.
3. *New flows and flow refreshes*—Constantly explore adding new flows that add value to the conversation. Add randomization decorators to keep the bot conversations fresh and engaging.

Building a bot is an ongoing process. In contrast to the mobile app update process, in most cases, making changes to your bot does not require users to reinstall it. Unless you are asking for new permissions (such as adding the permission to create channels or access additional users), you can update your bot in a transparent way. Because changing the conversation itself does not require users to update the bot, doing these improvements and tests on an ongoing basis is easy and frictionless.

[KEY TAKEAWAY]

Building a bot is an ongoing process. In contrast to the mobile app update process, in most cases, making changes to your bot does not require users to reinstall it.

For both B2C and B2B use cases, the design of a conversation is always evolving. As we come to understand our users better, we can evolve the design of our bot to make our users' lives more productive and fun.

[20]

To Infinity and Beyond—The Future of Bots

Prediction is very difficult, especially about the future.

—NIELS BOHR

We live in a super-dynamic software industry. Each technology revolution happens faster than the previous one, and each one brings new ideas, experiences, and opportunities to us all. In this chapter we will cover a few of the likely future trends in the bot industry and explore some of the upcoming opportunities and challenges.

Future Trends in Bot Platforms

Bot platforms such as Slack, Facebook Messenger, Kik, and others are constantly evolving. Similar to Android and iOS, each platform is taking a slightly different approach and thus providing a different feature set. They are all following the same general trends, however, which impact design in one way or another.

MORE INTERACTIVITY

The chat platforms realize that plain text limits bots to very specific use cases and limits users' ability to enter information in an easy and intuitive way. In a conversation I had with a lead engineer at Facebook, he stated it this way:

> The user can say the color of paint they would like to buy, but picking it from a color picker would probably be much easier.

There are two approaches to interactivity that we are going to see expanding in the near future:

Inline controls
: A growing set of controls, such as drop-downs, checkboxes, date pickers, and so on, are going to be presented inline in the conversation, just like the buttons, carousels, and Quick Replies we have seen in this book.

External pages
: Bots will be able to surface and expose to the users either webviews or templated windows to get their inputs.

Bot designers will be able, in the upcoming months and years, to provide a much richer experience and facilitate a lot more use cases with these types of rich inputs.

BETTER WAYS TO PRESENT INFORMATION

The chat platforms are also aware of the limitations bots have in presenting information to the user, whether it is geolocation information or data that can best be displayed in a dynamic chart. Here, we are going to see the same patterns of growing support for inline presentation elements that can be surfaced within a conversation, and external pages that can be popped up by the bot.

Here are a few examples:

- Charts, graphs, and tables—a way to present structured information
- Rich media—videos from different sources, third-party emulations
- Templates—tasks, galleries, events, reviews and ratings, dashboards, and more
- Maps—dynamic and closeable, with an overlay of information

[KEY TAKEAWAY]

Platforms will soon be providing designers with richer, more advanced ways to display and capture data.

DISCOVERY

Some platforms are more advanced than others in terms of providing users with an easy way to add a bot to a conversation, understanding how bots work, and initiating a workflow with a bot on an ongoing basis. But to be honest, all platforms currently suck at all of these right now. In the future we will see platforms provide more and more discovery mechanisms. For example:

- An in-chat-client way to search for and add a bot to a conversation
- The ability for a bot to add elements to the chat client's main real estate (like buttons or menu items, in the chat app's main interface) that will trigger workflows in the bot
- The ability to surface ratings, comments, and recommendations from users who have used the bots
- The ability to connect from the web straight into a chat with a bot (deep linking and referral support)
- In-chat-client bot suggestions—i.e., the ability of the chat platform to recommend useful bots, possibly based on the context of the conversation

MONETIZATION

Most platforms provide a very limited way to monetize bots at the moment. We are going to see a lot more ways to make money as bot platforms evolve, such as:

- Embedded billing support that enables the ability to charge for bot use on an ongoing basis, sale of in-bot virtual goods, and so forth
- Stronger integrations with Stripe, PayPal, and other payment providers
- The ability to pay more for a premium chat product that comes with premium bots preinstalled

At a high level, we are going to see rapid changes in chat platforms over the next few years, until bot design and the user paradigm solidify. Bot designers should continue to look for platform updates and trends. Some of the platforms have opened up their roadmaps to make that process easier.

SUPER BOT PLATFORMS OPENING UP

Google Home, Amazon Echo, Siri, and Cortana will all provide APIs for developers to plug in services, making these bots more powerful and more useful to the end user. This is not a bot feature per se, but rather the ability to extend the functionality of these "super bots" and to surface your services or products within these bot interfaces. There may be a slight problem of control over which service is offered or promoted first, with this determined by either the super bot or the end user.

As super bots are driven by big companies such as Google, Apple, and Amazon, I predict we will see a big push from these players to place these bots in strategic consumer devices such as phones and home devices and a drive for users to engage with and consume third-party services through these bots.

Future Trends in Bots

Bot designs themselves are evolving very rapidly, and we will see changes and advancements in many areas. We are in year one of bots as an industry, and we are learning quite a bit.

FROM NOTIFICATION TO INTERACTIVITY

Most bots that we've seen so far have been notification bots—bots that pump content into a chat application. In consumer use cases these have included things like weather notifications or news and entertainment notifications. In business use cases we have tended to see things like reports, reminders, and alert notifications.

While some of these notifications are useful use cases and they are usually really easy to implement, bot builders are realizing that users need to be engaged with and require interactive services that provide them with a lot of value in order to perceive bots as useful. We are still in the exploration stage, but we will be seeing a lot more use cases that implement real workflows and address major user needs and pains. Interactive bots, such as Kip, Statsbot, Growbot, Howdy, and Swelly, have performed really well and have even started to make money.

MULTIPLATFORM BOTS

"Write once, run everywhere" is one of the biggest promises in the high-tech industry, but it is also in many cases one of the biggest lies. What designers and developers usually experience in real life is "write once, run everywhere *slowly/uglily/at low quality*." We have seen this trend when trying to automatically port apps from iOS to Android and from the web to mobile. In addition, the use cases and audiences for chat platforms differ. Users on Kik are primarily teens, for example, while users on Facebook Messenger and Slack are adults, generally speaking.

There are two different approaches to providing bots across platforms:

- Provide the same user experience across all platforms by limiting the bot's functionality to the lowest common denominator. The upside of this approach is the ease of development and faster time to market.
- Separate logic from presentation and provide a different experience that is fine-tuned to suit the chat platform(s) the bot is on. The value here is in utilization of all of the goodness each platform has to offer, and a high-quality experience for the end user.

There have been some successful attempts at creating a bot that is exposed in several chat platforms, but they have all followed the same important principle—they took the time to *build a state-of-the-art user experience for their bot for each platform*. This has also been true for successful mobile and web app cross-platform experiences in the past. Because users are expecting the best-in-class experience on their particular platform, quick and easy is not your friend when it comes to cross-platform support.

[KEY TAKEAWAY]

When building a bot that will be exposed in multiple platforms, it is important to build a state-of-the-art implementation for each platform, rather than the lowest common denominator.

IDENTITY CONSOLIDATION

Handling identity with ease and grace is a challenge and a pain point for many designers and users. Having to register, reenter your preferences, and share sensitive information like credit card details with multiple bots is a big hurdle in user adoption and satisfaction. Additionally, bots expose services in platforms that already have a concept of user identity, requiring mapping of chat platform users to third-party service users. There are several approaches to solving these issues:

- Use the chat platform identity. Several platforms provide support for identity service at some level or another. Most are currently rather limited, but they should improve as time passes. The most advanced example of this approach is the Asian platform WeChat: it totally controls the user's identity and delegates that information to bots, and can even bill on behalf of the bot through an API.
- Use a third-party identity provider like Google, Facebook, or Okta. These are good solutions if you are already working with identity providers and delegated authentication.
- Perform a process to retrieve and map the chat platform users to your own users. These processes connect the platform's identity key (user ID) to your identity key.

This is still a murky area, and there is no clear winning approach when it comes to identity consolidation. I think platform-provided identity will be the way to go in the future, when platforms provide bot developers with a solid framework to manage user identities within and across bots.

AGILE CONVERSATION BASED ON USER SEGMENTATION AND SENTIMENT

Not all of us are the same. Users react differently to different prompts, different conversational styles, and different calls to action. This is also true of real-life conversations—young kids talk differently than teens, adults, and the elderly. Brazilians have a very different style than people from Japan. You would even have a different type of conversation with the same person depending on whether they were happy, sad, or annoyed.

Tools like Chatfuel have already started to provide conversational agility based on user segmentation, and tools like IBM Watson have demonstrated permutation in a conversation based on user sentiment analysis.

We are going to see more and more agility in conversations in the future, where bots adjust the tone, the style, and even the conversational flow based on the users and information gathered about them. Responsive conversation is what all humans are trained to do, and we will see this ability more and more in advanced bots.

DOMAIN-SPECIFIC BOTS VERSUS SUPER BOTS

There are two approaches that are emerging in the bot ecosystem: super bots and domain-specific bots. Super bots aim to expose multiple services using the same bot (one bot to rule them all), while domain-specific bots aim to expose a single service or product.

But what happens when there is a complex use case that involves multiple services? For example, planning a party, which involves ordering food, buying decorations, curating the music, inviting guests, and so forth? At the moment neither type of bot is able to deliver a good solution for these complex tasks, but in the future they will likely take very different approaches to solving this challenge.

Bot composition is a distributed approach to solving complex tasks—each bot takes a sliver of the problem and addresses it. Here is what the party-organizing conversation would look like:

> *User*: I want to organize a party again at my house next week!
>
> ***Catering-bot***: I will take care of the food ordering.
>
> ***Shopping-bot***: Ordering decorations :)
>
> ***Music-bot***: Should we use the same playlist we used last time?
>
> ***Friends-bot***: I'll invite the usual suspects.

The key benefit of this approach is that the user will have control over and a choice in the composition of the bots. The downside is a need for interoperability between the bots, and the overhead of setting up all these bots. If this approach prevails, we might see services that help the different bots communicate and work together.

Service orchestration is a centralized approach where a single super bot will take ownership of orchestrating all these services and provide a single interface for the user. Here is what this conversation might look like with a super bot:

> *User*: I want to organize a party again at my house next week!
>
> ***Super-bot***: I will take care of the food ordering. Also ordering decorations :)
>
> ***Super-bot***: Should we use the same playlist we used last time?
>
> ***Super-bot***: I'll invite the usual suspects.

The key benefit of this approach is the ease of use of having a single interface to work with. The downside might be reduced control and possible lack of transparency when it comes to who is actually providing the end services.

Google and Amazon are leading the charge with their super bots (Google Assistant and Amazon Alexa), while Slack, Facebook Messenger, Kik, and others are promoting the domain-specific bot approach. Only time will tell which will prove more successful in the long run.

Will Bots and AI Eat the World?

These are questions I hear at many conferences and read in many articles. Will bots replace apps? Will bots take people's jobs?

The short answer to these questions is, "Probably not in the foreseeable future." But let's try to answer each question in a bit more depth.

WILL BOTS REPLACE APPS?

Bots are a great hammer, but not everything's a nail. Many apps provide a rich and dedicated experience for a specific use case. I pity the fool who tries to implement a spreadsheet or a photo-editing tool or a first-person shooter game with a conversational interface. Bots are a great user experience to solve needs that can be addressed via a conversation. Bots are software taking the form of a personal assistant or a friend that interacts with us through a chat application. There are many use cases that bots will be better at than apps, but the opposite is also true.

In the next few years we are going to see a lot of services augmented through bots. This means bots will become yet another way to consume a service, alongside web and mobile apps. In conjunction, we will slowly start to see bot-only businesses emerge, as well as traditional services doubling down on their bot interfaces.

WILL BOTS TAKE PEOPLE'S JOBS?

Over the next few years we are also going to see bots augmenting people's work lives more and more, making them more productive and automating the boring parts of their jobs. An example will be support bots that can answer the common, easy, and repeatable questions, leaving the complex and high-value support tickets to human support agents.

There might be a day where AI is able to complete complex tasks like analyzing X-ray images or auditing content for legal use cases. But I think we are going to see more jobs being created as a result of AI and bots than jobs taken away by AI and bots.

Bots in Every Part of Our Lives

What we are going to see in the future is a lot more bots in our lives. At the last CES technology conference in Las Vegas, there were at least 10 home devices showcased that exposed conversational interfaces in one way or another. At work, we are all going to get our own personal assistants, just like we've (perhaps) always dreamed of. We will see many more bots functioning as sales assistants, HR assistants, legal and compliance aids, and more, all of them enabling people to be more productive. In our private lives, we will see brands exposing more and more services and products through chat; we will interact with software through voice in our cars and homes, and more and more commerce will be done through conversation in our favorite platforms.

[KEY TAKEAWAY]

We are going to see bots everywhere, making our lives better—conversational interfaces are the future of software.

[Index]

C

J

K

L

S

T

U

V

W

X

Y

Z

About the Author

Amir Shevat, head of developer relations in Slack, works with bot developers and designers. Previously, he managed Google's Startup outreach program, helping developers around the world design and build better products. Amir has also created a product design course in Udacity, teaching product managers, designers, and developers how to build products users love.

Colophon

The animal on the cover of *Designing Bots* is a Siberian Husky, a medium size breed of dog that was originally bred to assist the native Chukchi people with hunting and transporation. The Siberian Husky can be distinguished by its thick fur, large ears, and distinctive coloring.

The coat of the Siberian Husky consists of two layers: the undercoat and the topcoat. These two layers protect the dogs from the extremely harsh weather conditions, allowing them to withstand temperatures as low as –60 °C. Their coat can also reflect sunlight during the long daylight hours of the Arctic summer.

Siberian Huskies are often used to pull sleds, one of the only ways to travel during winter in remote regions of Russia and Alaska. The breed became popular after the story of Gunnar Kaasen and his Siberian Husky Balto delivered medical supplies by sled to the city of Nome, Alaska in 1925.

Many of the animals on O'Reilly covers are endangered; all of them are important to the world. To learn more about how you can help, go to *animals.oreilly.com*.

The cover image is a color illustration by Karen Montgomery. The cover fonts are URW Typewriter and Guardian Sans. The text font is Scala; and the heading font is Gotham.

CPSIA information can be obtained
at www.ICGtesting.com
Printed in the USA
BVOW07s1514060817
491298BV00005B/25/P